# Legal competence in environmental health

# Also available from E & FN Spon

**Clay's Handbook of Environmental Health**
17th edition
W H Bassett

**Environmental Health Procedures**
4th edition
W H Bassett

**European Union Health and Safety Legislation - Volumes 1 to 6**
A C Neal and F B Wright

**European Union Public Health Legislation**
A C Neal and F B Wright

**Fundamentals of Noise and Vibration**
F J Fahy and J G Walker

**Homes and Health**
B Ineichen

**Occupational Health and Safety Law in the UK and Ireland on CD-ROM**
A C Neal and F B Wright

**Occupational Health and Safety Law in the European Union on CD-ROM**
A C Neal and F B Wright

**Statistical Methods in Environmental Health**
J C G Pearson and A Turton

**Sick Building Syndrome**
J Rostron

**Water Quality Assessments**
D Chapman

*For more information about these and other titles please contact:*
The Marketing Department, E & FN Spon, 2-6 Boundary Row, London, SE1 8HN. Tel: 0171 865 0066

# Legal competence in environmental health

## Terence Moran

Leeds Metropolitan University,
Leeds, UK

**E & FN SPON**
An Imprint of Thomson Professional

London · Weinheim · New York · Tokyo · Melbourne · Madras

**Published by E & FN Spon, an imprint of Chapman & Hall, 2–6 Boundary Row, London SE1 8HN, UK**

Chapman & Hall, 2–6 Boundary Row, London SE1 8HN, UK

Chapman & Hall GmbH, Pappelallee 3, 69469 Weinheim, Germany

Chapman & Hall USA, 115 Fifth Avenue, New York, NY 10003, USA

Chapman & Hall Japan, ITP-Japan, Kyowa Building, 3F, 2-2-1 Hirakawacho, Chiyoda-ku, Tokyo 102, Japan

Chapman & Hall Australia, 102 Dodds Street, South Melbourne, Victoria 3205, Australia

Chapman & Hall India, R. Seshadri, 32 Second Main Road, CIT East, Madras 600 035, India

First edition 1997

© 1997 Terence Moran

Typeset in 10/12pt Palatino by Saxon Graphics Ltd, Derby, UK
Printed in Great Britain by TJ International Ltd, Padstow, Cornwall

ISBN 0 419 23000 9

 Printed on permanent acid-free text paper, manufactured in accordance with ANSI/NISO Z39.48-1992 and ANSI-NISO Z39.48-1984 (Permanence of Paper).

# Contents

# Preface

In the case of *Welton v North Cornwall DC* (1996) it was found that, where an Environmental Health Officer negligently required the owner of food premises to undertake works which were unnecessary to secure compliance with the Food Safety Act 1990 and the owner incurred substantial and unnecessary expense in carrying out the works, the local authority was under a common law duty of care to the owner and liable for the £34,000 in damages.

What this case illustrates are the two themes which have informed the creation and development of this book. Growing public awareness of Environmental Health issues, coupled with necessity and expediency, has promoted a wide range of relevant and complex legislation over recent years. This is against a background of enhanced understanding by the general public of their legal rights. These processes continue to have substantial implications for those engaged in Environmental Health both within and outside local government and for the professional competencies required. The strength of the Environmental Health profession has long been held to rest on having a broad knowledge base across a wide range of disciplines, thereby promoting a holistic approach to the solution of environmental problems. Valuable as this may be, there is an increasing need for high levels of specialist technical expertise. This has long been recognised with regard to matters of Food Safety, Occupational Health etc. But this has not been the case with regard to the law; perhaps now it is.

There exist a number of perfectly sound books which describe what may be called 'Environmental Health Law'. However, in common with many books on the law, these have a tendency to state what the law is but do not then go on to describe how it works. The nearest parallel I could draw is to that of DIY car maintenance. Buying a book on cars or even several books on cars may tell you a great deal about cars, their history, design, manufacture, engine etc. But it will not tell you how one works nor how you may fix it. For this a car manual is required. A good manual will not simply describe the parts of the car but will tell how they work and how the parts relate to each other. To explain a car in terms of how it is made and what it is made of will not tell you how to use that car, how it may be disassembled or what tools you may need to fully benefit from the vehicle.

So it is that to simply explain the law in terms of what it says is not enough. An Environmental Health Officer will need to know not only what the law is, but also how to use it. That is what this book is meant to assist with. It examines those practical legal skills that are essential to the proper discharging of the Environmental Health Officer's duties. Hopefully it will offer answers, not only to the question, what is the law?, but also to the question: how do I use that law? This book is not an expo-sition of all Environmental Health law, it is designed to indicate how that law may be applied, emphasising those skills and attributes which are essential if the theory of law is to be put into professional practice.

What follows encompasses an examination of those aspects of the work of the Environmental Health Officer which, though regulated by legal principles, are not to be found in a single statute or case. I would suggest, the not very startling proposition that Environmental Health Officers must not simply 'know the law' but must understand its imple-mentation. This requires them to possess and display competencies in legal draftsmanship, obtaining evidence, selection of legal remedies, advocacy etc. The aim of the text is thus to concentrate less on the opera-tion or application of a particular statute or statutory provision and more on an examination and emphasis of those skills vital when operating in such a broad, though legally based, discipline. The approach is to look at the generic legal skills vital to a professional EHO.

Environmental Health Officers have always operated within a 'legal' environment. Yet I feel it fair to say that they have not been subject to the full rigour of the law. However now, more than ever, the work of the Environmental Health Officer is subject to scrutiny and in their use of the law they cannot afford to be found wanting.

THE CONTEXT OF ENFORCEMENT ON ENVIRONMENTAL HEALTH PRACTICE

There have always been deficiencies in the English common law as a means of protecting the health of the individual or the quality of the environment. The traditional approach was not about pre-emption but largely about reparation for damage caused, with all the attendant diffi-culties of locus, causation etc. At best the common law took a very lim-ited view on environmental health matters and was always primarily reactive. The approach was, and is, centred on the individual; dealing with issues only to the extent that they impact on the rights of individuals as individuals, not as members of a community.

It is difficult to argue that such an approach could have been held to be effective when dealing with threats affecting commons goods. The historical challenge was thus one of devising rules to govern commons and to address issues of matter to the community. The traditional approach by lawyers when confronted with new problems has always been to attempt to deal with them using pre-existing remedies applied via pre-existing principles, not to go back to the drawing board.[1] Yet it must be self-evident that sometimes new problems will demand new solutions. One new solution in the nineteenth century was the Sanitary Inspector. As the Environmental Health Officer of today however, they are no longer a 'new solution' but instead often find themselves as part of an 'old' solution confronting new problems. And these are not simply new problems of pollution, poor food or bad housing. Instead they confront problems of resourcing, politics, technology and a changing social expectation. If they are not themselves to be replaced by a 'new solution' they must change. Old attitudes are required to be re-examined in the light of new imperatives, old practices are in need of replacement. The decisions Environmental Health Officers make about how to deal with issues are no longer the same. How they must interpret and use the law has changed. They have been identified in the past as the bridge between the government's decision to intervene and protect the environment and the impact of this intervention upon both the environment and the regulated. But if this is indeed what they are, then the design of that bridge is not fixed.

The Environmental Health Officer's interpretation of his own function is under scrutiny; how do they see themselves and is that how others see them? Are they a police force primarily concerned with the apprehension and subsequent punishment of offenders or are they there to achieve, via a variety of mechanisms, compliance with relevant standards and thereby protect public health? Are they environmental managers or educators, facilitators or a source of constraint?

The jury is still out on the answers to these questions, but they must be answered for they will certainly suggest the best means of operating within an evolving legal framework.

HOW TO USE THIS BOOK

To gain the best advantage out of this book I would advise that it be used as a source of reference rather than cover to cover reading, though, hopefully, it will offer guidance on good practice in certain areas.

Generally I have tried to structure the book as if following a single case, asking questions such as: What is the basic law? How may I get into the premises to begin an investigation? What are the relevant considerations in undertaking an investigation? What shall I do as a result of the investigation? Certainly it may be possible to examine specific statutory provisions using this book, but overall the aim would be to use the text in tandem with, for example, the text of an Act or set of regulations. In this way the latter may be used to provide the detail and the former the interpretative skills.

## NOTES

1 Mayda J (1978), Penal Protection of the Environment, *Am J'nl of Comparative Law*, Vol 26, 471.

# Sources of law                                      1

The aim of this book is to assist Environmental Health Officers to understand the operation of English law and to navigate through some of its complexities. To do this it is first necessary to examine the basic elements and thus establish an understanding of the origins and sources of English law. There is no single source for all of the laws to be found in this country, however the main sources of law are here taken to be:

- Legislation
- Judicial precedent
- European law

It is accepted that Custom is often and correctly quoted as an additional source of law, however for those engaged in the Environmental Health profession it is not a significant source and is of limited importance. It will thus not be considered further here. For those that have been mentioned each will be considered in turn as a source of law influential in the work of the Environmental Health Officer.

## 1.1 LEGISLATION

### 1.1.1 ACTS OF PARLIAMENT

Parliament has a number of functions. It is a forum for the redress of individual grievances. It is a place for the scrutiny and criticism of government policy and of the executive. Arguably however the primary role of Parliament is as a law-making body; it is a legislature and there is little doubt that in our modern world the most important single source of law is legislation. Included within the term legislation are both primary legislation – Acts of Parliament and delegated legislation – Regulations, orders etc. For the moment it is only the former which are to be considered. Parliament, as the supreme lawmaking body in this country passes

Acts. These then become the law of the land. The passing of an Act may make new law or alter or repeal existing law. It is important to remember that this ability to alter existing law applies, not simply to existing statutory law, but also to the extant common law.

### 1.1.2 THE LEGISLATIVE PROCEDURE

For an Act to be passed it must first have gone through the appropriate procedure moving from a bill, the proposed form of the Act but having no legislative effect, to the Act proper. The important public bills, which will result in Acts having general effect throughout the country, will have their origin in Cabinet. It is the Cabinet which will, via its legislation committee, consider departmental proposals for government bills. A bill is usually introduced into the Houses of Parliament by a ministerial sponsor on behalf of the government. The passage of a bill will usually start with its introduction into the House of Commons, though in fact it may be initiated in either the Commons or the Lords. A Notice of Presentation is placed on the relevant day's order paper and the title of the bill is read by the Clerk. This constitutes the 'first reading' and the bill is then said to be received for first time. The bill will then be printed and published by HMSO. A second reading date will also be set. The precise time may vary but approximately two weeks after this first reading the second reading will take place.

### 1.1.2.1 Second Reading

This stage in the proceedings provides an opportunity for debate on the principles contained within the bill. This debate provides the government with its first opportunity to speak on the bill and it also provides a means by which the attention of the whole House can be drawn to and focused on the bill. Where proposals contained in the bill are opposed then both supporters and opponents will be allowed to speak.

When the proposals contained within the bill are opposed a vote will be taken. If the bill is not approved at this stage it cannot proceed to the next stage. If the bill is approved it is allowed to proceed and it will then move on to the next stage in the procedure which is the 'Committee Stage'.

### 1.1.2.2 Committee Stage

Most public bills are automatically referred to a Standing Committee. The numbers can and do vary on these committees but may usually be 20 to 40 MPs. It is at this stage that the bill is examined clause by clause and it is this process which provides the only real opportunity to consider the line by line detail of the bill.

There are alternatives to a Standing Committee and on occasion a bill will go to a Committee of the Whole House. This, for example, may be when the bill involves matters of constitutional significance. A further alternative is a Select Committee. Such a committee may, as part of its activities, call individuals to give evidence before it. After the Committee Stage comes the 'Report Stage'.

### 1.1.2.3 Report Stage

The chairman of the committee reports back on the bill to the House, indicating any amendments which have come out of the deliberations undertaken at the Committee Stage. This Report Stage provides a further opportunity for the bill to be considered, now in its revised form. Only the amendments inserted at the Committee Stage are subject to debate and further alterations may be made. Between one and two weeks later the amended bill may go on to the next stage – the 'Third Reading'.

### 1.1.2.4 Third Reading

This is a one-day debate on the bill as it has been amended. At the conclusion of this a vote is taken on the bill and the motion, if carried, means that the Commons' stage of proceedings, if that is where they were started, is then concluded. The bill is now at the 'Transfer Stage', where it is transferred to the House of Lords to go through a similar process there, as it has just gone through in the Commons. In the Lords the stages of progression are the same as in the Commons except that the committee is always of the whole House.

After third reading in the Lords the bill has reached the 'Messages Stage' and it is then returned back to the Commons complete with the Lords' amendments. The House of Commons must consider the Lords' amendments and if they cannot agree them then the bill is returned. If the Lords and Commons cannot reach an agreement the bill is lost for that session, though it may be introduced in the next session of Parliament. When the final form of the bill is agreed upon the bill will receive the Royal Assent and be published as an Act within two weeks.

This is obviously a complicated and time-consuming process, not readily accessible to most of the public. Yet the significance of the procedure should not be underestimated, not simply because it is a means by which law is made but also because it offers the opportunity to understand how to interpret that law.

## 1.2 CONTENTS OF AN ACT OF PARLIAMENT

An Act of Parliament is, for the Environmental Health Officer, one of the most important primary sources of information. It is of course right that busy individuals should refer to books and encyclopaedias for advice on interpretation and application of the law, but this should never replace consultation of the primary document, the Act itself. To be able to maximise the use of an Act of Parliament it is vital that the Environmental Health Officer understands how Acts are organised and how they 'work'. Certainly they are a collection of words on paper which have legal effect, but to think that is all an Act is is to fail to appreciate the thing fully and, thereby, to run the risk of failing to interpret and thus apply it correctly. An Act is Parliament's best attempt to regulate activity in a given area. Sometimes the Act succeeds admirably in doing this, sometimes it does not. This may sometimes be because of bad draftsmanship and sometimes it may simply be because of the limits of language. An Act has a form and a content. It has a history and an aim; these must be appreciated by the Environmental Health Officer if he is to interpret and apply the law correctly.

### 1.2.1 LAYOUT OF AN ACT

An Act of Parliament begins with its 'short title'. In 1896 The Short Titles Act of that year provided that each Act of Parliament should have a short title by which it may more easily be known. This is then the title by which the Act will commonly be known and referred to; for example, The Housing Act 1985, The Health and Safety at Work etc. Act 1974, the Environmental Protection Act 1990. This is the first element in the layout of the Act.

There will then follow the official citation for the Act. This is discussed in more detail later. After this official citation will come the 'long title' of the Act. This in fact looks most unlike a title but is of significance in understanding the purpose of the Act. The long title will begin with the words 'An Act to ... ' Its purpose is to state the aims of the Act and indicate its application. For example the Food Safety Act 1990 states, as its long title, that:

> it is an Act to make new provisions in place of the Food Act 1984 (except Parts III and V), the Food and Drugs (Scotland) Act 1956 and certain other enactments relating to food; to amend Parts III and V of the said Act of 1984 and Part I of the Food and Environmental Protection Act 1985; and for connected purposes.

After the long title is to be found the date of Royal Assent. Every statute since 1793 has carried this and it will often, but may not always, be the date on which the Act comes into effect. After the date of the Royal Assent will come the Words of Enactment or Enacting Formula. These words are always the same and begin with the words 'Be it enacted ...'

### 1.2.1.1 Parts of an Act

An Act will then be found to be divided into Parts. These parts are numbered using roman numerals e.g. Part I, Part IV. Each of the Parts of the Act will deal with a different subject or aspect of the activities or matters regulated by the Act. For example Part I of the Environmental Protection Act 1990 deals with Integrated Air Pollution Control and Air Pollution Control by Local Authorities, Part II deals with Waste on Land and Part III deals with Statutory Nuisances and Clean Air. An Act will have as many Parts as are needed.

The Parts of an Act are then subdivided into Sections. These sections are themselves numbered consecutively and the number is presented in bold type in the body of the Act. Sections are given the abbreviation 's'. The sections themselves are further divided into sub-sections. These are also numbered sequentially for the section but the numbers are contained within round brackets ( ) and are abbreviated to ss. The sub-sections may themselves be further divided into paragraphs. These paragraphs are not numbered but are instead given letters in alphabetical order. These are again bracketed and are in lower case. Thus the part of the Health and Safety at Work etc. Act 1974 which deals with the duty of an employer to his employee with regard to plant and systems of work is to be found in Part I, section 2, sub-section 2, paragraph a. This may be more easily recorded as s.2(2)(a) of the Health and Safety at Work etc. Act 1974 . Similarly the part of the Environmental Protection Act 1990 which identifies premises as capable of being a statutory nuisance is Part III, section 79, sub-section 1, paragraph a or s.79(1)(a).

### 1.2.1.2 Side Notes

Also included near to the sections are the side or marginal notes, which give a brief summary of what the section is about. For example the side note to s.8 of the Food Safety Act 1990 states that it deals with 'selling food not complying with food safety requirements' and the side note to s.79 of the Environmental Protection Act 1990 records that it deals with statutory nuisance and inspections therefor. Side notes are extremely useful in identifying what a section is dealing with without the need to read the whole section first. Also the side note is of aid in indicating the mischief which the Act is aiming to address. However it should not be used as an aid in determining the scope of the Act. In the case of *R v Kent*

[1977] the Court of Appeal had to consider whether the appellant had with him a firearm contrary to s.18(1) of the Firearms Act 1968. The side note to the section said 'Carrying firearm with criminal intent'. The court found that this note was of no assistance in establishing the scope of the section in question. Lord Scarman stated that 'it used to be thought that one could have no regard to marginal notes in studying the meaning of an Act of Parliament, but the House of Lords made it clear in *DPP v Schildkamp* [1971] that regard may be had to a marginal note, not to interpret the Act of Parliament, but as an indication of the mischief with which the Act is dealing.'

### 1.2.1.3 Definitions Section

A most useful part of an Act is the definitions section. Here can be found a definition for many of the more commonly occurring words or phrases used in the Act. For the interpretation of such words one should not be tempted to rely on a dictionary definition. The meaning attached to something in a dictionary may not be that attached to it in the Act. Also it is important to note that the definition applies only to the word or phrase used within that Act, it is not intended to be applied generally across all other statutes. So, by way of example, in the case of the Environmental Protection Act 1990, there is more than one definitions section; s.1 is the relevant section for Part I of the Act, s.29 for Part II, s.79 for Part III and so on. In the Food Safety Act 1990 and the Health and Safety at Work etc. Act 1974 the interpretation section happens to be the same: s.53. The significance of the application of the interpretation section should not be underestimated when applying the provisions of the Acts. In *Westminster City Council v Select Management Ltd* [1984], improvement notices were issued by Westminster City Council, under s.21 of the Health and Safety at Work etc. Act 1974. The notices related to lifts and an electrical installation in a block of flats. The matters covered concerned an alleged breach of s. 4 Health and Safety at Work etc. Act 1974. This section deals with the general duty of an owner-occupier and others having control of non-domestic premises, when those premises are made available, as a place of work, to persons other than their employees. Select Management Ltd managed the flats. Workmen, who were not the employees of Select, occasionally carried out maintenance on the lifts and other appliances. Select claimed that the local authority did not have the power to issue notices in respect of only part of a building. The lifts etc. were part of a building and that building was domestic. As s.4 of the Health and Safety at Work etc. Act 1974 does not apply to domestic premises there could thus be no power available to the authority to issue the notices. The court rejected this argument and accepted that put by the local authority. Westminster contended that s.53 of the Act, the definitions section, defined domestic premises as including any appurtenance of a private dwelling which is not used in common by the occupants of any such dwelling. Thus any parts of a building which were used in common, such as lifts, were non-domestic premises. The lifts, being outside the definition contained in s.53, were therefore regulated by the Act and improvement notices could properly be issued.

### 1.2.1.4 Commencement

The last section of an Act will deal with the Short Title, commencement and extent of application of the Act. This section will state what the Short Title of the Act is and indicate when the Act, or parts thereof, will come into force. This need not and often will not be on the date given for the Royal Assent. Indeed it is common now to find a provision in a statute which gives the government a discretion to bring parts of the Act into effect at some later time by means of 'commencement orders'. These enable the government to give legal effect to provisions contained in an Act when and how it pleases.

It is an assumption of statutory interpretation that an Act will apply to the whole of the UK unless it is indicated to the contrary in the Act.

The remainder of the Act will be made up of schedules which provide additional information to that contained in earlier parts of the Act. One of these schedules will be a repeals schedule indicating the extent to which the current Act repeals earlier legislation. For example Schedule 24 of the Environment Act 1995 deals with repeals and revocations resulting from that Act. Schedules are referenced only by paragraph and sub-paragraph.

### 1.2.2 CITATION OF ACTS OF PARLIAMENT

Before 1850 the citation of a statute was by its long title. This was a cumbersome means of referencing, so much so that now Public Acts are normally referred to by their short title, though they do in fact have a more complete reference. For example the Environmental Protection Act 1990 has an official reference or formal citation of: **1990 c.43.** This reference, regulated by the Acts of Parliament Numbering and Citation Act 1962 means that it was the 43rd Act to be passed in the year 1990. The chapter number (c.) is assigned by reference to the year in which the Act was passed.

Acts passed before the 1962 Act have a different form of citation which made use of the 'regnal year or years' during which the Parliamentary session in which the Act was passed took place. Thus the Public Health

Act 1936 may be cited as: **26 Geo.5 & 1 Edw.8 c.49** This reference indicates that the Act was passed during the 26th year of the reign of George V and the first year of the reign of Edward VIII.

## 1.3 DELEGATED LEGISLATION

The world is becoming an ever more complex place. The depth and breadth of knowledge acquired reveals ever more complex problems and the law is required to assist in addressing and solving such problems. The complexity of the law in the area of Environmental Health makes it a practical impossibility that all of the detail necessary for effective regulation could be contained in a single Act of Parliament. As long ago as 1932 it was identified, by the Committee on Ministers Powers (Cmnd. 4060 (1932)), that an Act of Parliament was not particularly well suited to respond to these needs. Its creation and change is a slow business; the procedures involved in its creation are inappropriate when dealing with matters of great complexity. It is held to be rigid and inflexible and unable to provide a speedy remedy when the situation demands it. It is therefore very common to see an Act of Parliament drafted so as to provide the framework onto which more detailed provisions can be grafted; the skeleton onto which flesh may be put. Such an Act may provide the broad general framework of the law in a particular area but no more. It leaves it to others, to whom a law-making power has been delegated, to provide the necessary detail. Those empowered to make such rules and regulations, for example a Secretary of State, are then able to create legislative instruments for a specific purpose or purposes. These instruments have the same legal force and effect as the Act under which they are made, yet they are less rigid and take much less time to enact. Delegated legislation is a form of law with which all Environmental Health Officers must be familiar as it is found in all areas of their professional activity. For example s.15 of the Health and Safety at Work etc. Act 1974 deals with health and safety regulations. The section empowers the Secretary of State to make health and safety regulations for any of the general purposes of Part I of the Act. The section describes the ambit of those regulations while Schedule 3 of the Act gives details of the particular purposes for which regulations may be made.

Regulations made in this way:

(a) may repeal or modify any of the existing statutory provisions;
(b) may exclude or modify in relation to any specified class of case any of the provisions of sections 2 to 9 or any of the existing statutory provisions;

(c) may make a specified authority or class of authorities responsible, to such extent as may be specified, for the enforcement of any of the relevant statutory provisions.

Thus it was possible, by the use of delegated legislation, in fact the Health and Safety (Enforcing Authority) Regulations 1989, to increase the enforcement responsibilities of local authorities.

The power to make such regulations allows the general provisions, which are to be found throughout the Act, to be augmented by the creation of more detailed requirements designed to add to the general provisions of the Act more specific duties. For example the general duty of an employer to his employee to be found under s.2 of the Health and Safety at Work etc. Act 1974, whilst without doubt extending to protection against noise-induced hearing loss, is so general in its application that, with regard to exposure to excessive noise it is hard to be sure quite what the obligations on an employer are. However an examination of the Noise At Work Regulations 1989, which were made under the Act, and apply to all situations where the Health and Safety at Work Act applies, reveals a number of new and quite specific duties on employers with regard to risks to hearing in the workplace. These include:

- a requirement to make, review and record noise assessments
- the requirement to reduce the risk of hearing damage to employees
- the requirement to reduce noise exposure
- the requirement to provide personal hearing protection
- the creation of ear protection zones
- use and maintenance of equipment
- and the duty to provide information, instruction and training.

Clearly the regulations, whilst in no sense replacing the duties to be found in s. 2 of the Health and Safety at Work etc. Act 1974, add to those duties by the creation of obligations which are much more specific and detailed, with regard to the particular hazard of excessive noise, than those to be found in the Act. For those subject to control the regulations make it much easier to know when the statutory obligation with regard to noise is being met.

It is not just in the area of occupational safety that regulations add to the detail of the legislation enforced. Section 369 of the Housing Act 1985 enables the Secretary of State to make regulations for the standards of management in Houses in Multiple Occupation. The Housing (Management of Houses in Multiple Occupation) Regulations 1990 elaborate on the standards of management to be applied in a house in multiple occupation. Matters covered by the Regulations include:

1. Good order and continuity of water supplies and drainage

2. Repair and continuity of gas and electric supplies and installations for their use
3. Lighting
4. Repair, condition and cleanliness of the common parts of the house
5. Repair of windows and other means of ventilation
6. Repair and freedom from obstruction of fire escapes
7. General precautions to protect occupiers from injury as a result of the condition of the house.

For those charged with enforcement, questions of compliance or non-compliance are much more easily answered.

### 1.3.1 TYPES OF DELEGATED LEGISLATION

There are four main types of delegated legislation which those engaged in Environmental Health may encounter.

#### 1.3.1.1 Statutory Instruments

These are made under the procedure found in the Statutory Instruments Act 1946 and are the most commonly encountered form of delegated legislation for the Environmental Health Officer. Section 1 of the 1946 Act defines a Statutory Instrument and states that where, by any Act passed after 1 January 1948, a power to make, confirm or approve orders, rules, regulations or other subordinate legislation is conferred on the Queen in Council or on any Minister of the Crown then if the power is expressed

1. in the case of a power conferred on the Queen, to be exercisable by Order in Council
2. in the case of a power conferred on a Minister of the Crown, to be exercisable by statutory instrument, any document exercised by that power shall be known as a 'Statutory Instrument'.

The procedural requirements for the making of this form of legislative instrument are regulated by the 1946 Act and the relevant Act under which the power is delegated. For example under the Health and Safety at Work etc. Act 1974, s.50 provides for there to be consultation between the Secretary of State and the Health and Safety Commission and for there to be further consultation with such other bodies as appear to be appropriate, prior to the making of any regulations.

The procedures for the approval of a statutory instrument may make it subject to one of two Parliamentary procedures: the Affirmative Parliamentary Procedure or the Negative Parliamentary Procedure. In the less common 'Affirmative' procedure the statutory instrument is laid in draft before Parliament. The laying of a document means the delivery of a copy of it to the Votes and Proceedings Office of the Commons and the Office of the Clerk to the Parliaments. No amendment to the proposal is normally possible in this procedure and the instrument will have no effect until it is approved by an affirmative resolution of the House. The second and more common Negative Parliamentary Procedure, contained in s.5 of the Statutory Instruments Act 1946, requires that the statutory instrument is first laid before Parliament. If either House, within a period of 40 days, beginning with the day on which a copy of the proposal was laid, resolves that an address be presented to Her Majesty that the instrument be annulled (this is commonly called a prayer), then no further proceedings shall be taken and Her Majesty may revoke the instrument. If no such prayer is moved the instrument will move on to take effect.

There also exists the Joint Select Committee on Statutory Instruments. This committee may scrutinise statutory instruments with a view to deciding whether the attention of Parliament should be drawn to a statutory instrument on any of the following grounds.

1. It imposes charges on public revenue.
2. It seems to be immunised against challenge in the courts.
3. It purports to have retrospective effect in the absence of express authority from the parent Act.
4. There appears unjustified delay in publication or laying before Parliament.
5. It is essential for the statutory instrument to come into effect before being laid and there has been unjustified delay in telling the Speaker of this and the reasons.
6. The statutory instrument may be *ultra vires*.
7. The statutory instrument requires elucidation.
8. The draftsmanship is defective.

Before reporting to the House the Department sponsoring the statutory instrument is notified and has a chance to reply to the Committee's findings.

A statutory instrument will come into effect when it is made unless some other date is specified. Section 3(2) of the Statutory Instruments Act 1946 provides a defence to criminal proceedings for contravening a statutory instrument. This section stipulates that, in any proceedings against any person for an offence consisting of a contravention of a statutory instrument, it shall be a defence to prove that the instrument had not been issued by Her Majesty's Stationery Office at the date of the alleged contravention. That is unless it is proved that at that date reasonable steps had been taken for the purpose of bringing the purport of the instrument to the notice of the public, or of persons likely to be affected by it, or of the person so charged. This, though potentially a problem an Environmental Health Officer may encounter, is in practice highly unlikely. The actual name given to a statutory instrument: – Regulation, Order, Direction – is often merely a function of a particular departmental practice.

### 1.3.1.2 Orders in Council

Powers are conferred on the Queen in Council. As a matter of fact these powers are exercised by the Cabinet. Such Orders may be made by exercising the Royal Prerogative but are more commonly made under a relevant statutory power. If this is so they will be statutory instruments and the provisions of the 1946 Act will apply: e.g. Control of Noise (Code of Practice for Construction and Open Sites) Order 1984 SI 1984 No 1992.

### 1.3.1.3 Bye-laws

Section 235 of the Local Government Act 1972 deals with the power of a council to make bye-laws for good rule and government and the suppression of nuisances. The section provides a power whereby a council may make bye-laws to achieve the stated purpose for the whole or any part of the district or borough. The Secretary of State is identified by the section as the confirming authority in relation to bye-laws made under this Act, and the section limits the power to make bye-laws to those situations which are not already covered by some other provision made under another enactment. In other words bye-laws ought not to be passed to regulate matters already regulated by Parliament. The procedure for the making of such bye-laws is contained within s.236 of the Act. Section 236(3) requires that bye-laws be made under the common seal of the authority and should not have effect until confirmed by the confirming authority. At least one month before application for confirmation, notice of an intention to apply for confirmation must be given in the local newspapers and for the period of a month before application for confirmation a copy of the bye-law must be made available, without charge, for public inspection. A copy of the bye-law, when confirmed, must be similarly made available for public inspection.

### 1.3.1.4 Rules of the Supreme Court

Under the Supreme Court Act 1981, the Rules Committee may make rules concerning practice and procedure in the Court of Appeal, the High Court and the Crown Court. Similarly under the County Courts Act 1984 rules may be made for the County Court. Probably the only time an Environmental Health Officer is likely to come into contact with these rules is if he or she becomes involved in a matter subject to judicial review. The essential procedural requirements for an application for judicial review are to be found in the Rules of the Supreme Court, Order 53, as re-enacted in s.31 of the Supreme Court Act 1981.

### 1.3.2 LAYOUT OF A STATUTORY INSTRUMENT

The first element of the instrument encountered is its title and the year and number of the instrument. There will then follow the dates the instrument was made, laid before Parliament and comes into force. This will then be followed by a recital naming the person exercising the delegated power and the relevant enabling provisions.

In the Statutory Nuisance (Appeals) Regulations 1995 (which revoked the earlier Statutory Nuisance (Appeals) Regulations 1990 and Statutory Nuisance (Appeals) (Amendment) Regulations 1990) this reads as follows:

> The Secretary of State for the Environment, as respects England, and the Secretary of State for Wales, as respects Wales, in exercise of the powers conferred upon them by paragraph 1(4) of Schedule 3 to the Environmental Protection Act 1990 and of all other powers enabling them in that behalf, hereby make the following Regulations.

There will then follow the main body of the instrument. The individual elements or regulations are numbered and referred to as 'Regulations' not sections. It is thus wrong to speak of section 1 of the Noise at Work Regulations 1989; the correct notation is Regulation 1. In full regulations are divided in the following way:

- Part
- Regulation
- Paragraph
- Sub-paragraph.

The instrument may contain schedules as for an Act of Parliament. The first regulation or regulations will normally deal with citation, commencement and interpretation and later there will an indication as to who signed the instrument and an 'explanatory note'.

### 1.3.3 CITATION OF DELEGATED LEGISLATION

Though delegated legislation is commonly known by the equivalent of a short title the correct citation of these instruments is a basic skill for the

Environmental Health Officer. The citation of such legislative instruments is controlled by the Statutory Instruments Act 1946. This Act, which came into effect on 1 January 1948, repealed previous provisions whereby before 1948 these forms of legislation were known as Statutory Rules and Orders. Knowledge of their former name helps to understand their method of citation. Using the example of The Cremation Regulations 1930, the former method of citation used for these rules and orders was: **The Cremation Regulations 1930 S.R. & O. 1930 no. 1016 or S.R. & O. 1930/1016.**

Since 1948 such legislative instruments have, however, been known as Statutory Instruments – SI's. Section 2 of the Statutory Instruments Act 1946 deals with the numbering, printing, publication and citation of such instruments. The Act provides that immediately after the making of any statutory instrument, it must be sent to the Queen's Printer and numbered in accordance with regulations made under the Act. The relevant regulations, made under s.8 of the Act, are the Statutory Instruments Regulations 1947. These require that instruments be allocated to the series of the calendar year in which they are made and be numbered in that series consecutively and as nearly as may be in the order in which they are received.

Thus the full citation for the 1947 Regulations is: **The Statutory Instruments Regulations 1947 SI 1948 no 1.**

Any statutory instrument may be cited by the number given to it in accordance with the provisions of this section, and the calendar year, i.e. **SI 1948/1.**

For the Environmental Protection (Prescribed Processes and Substances) Regulations 1991 the reference is: **SI 1991/472**

For the Housing (Houses in Multiple Occupation) Management Regulations 1990 the reference is: **SI 1990/ 830**

## 1.4 FINDING LEGISLATION

However good the book or encyclopaedia being used there will come a time when reference to the original Act or Regulations is necessary. This can prove to be much easier said than done. There are a number of sources which may be used to find legislation. All are to be found in a dedicated law library but will rarely be available elsewhere.

### 1.4.1 CURRENT LAW LEGISLATION CITATOR

Here the Acts are arranged chronologically and the contents cover amendments, repeals, statutory instruments and significant cases since 1947.

1.4.2 HALSBURY'S STATUTES

This contains in alphabetical order, by subject, current amended legisla-tion. It contains references to statutory instruments and cases and is cross-referenced to:

Halsbury's Statutory Instruments
Halsbury's Laws of England
The Digest
All England Law Reports.

If a statute being sought has been repealed then Halsbury's statutes will be of little help as it does not include repealed legislation. However it has a number of other purposes. It will allow the citation of an Act to be checked and this can be used to find a copy of the Act in question. It can also be used to find out which sections of that Act remain in force. Because the Acts contained are annotated it is these annotations which make it possible to find key cases relating to that Act and to find relevant statutory instruments made thereunder. By using Halsbury's it is also possible to trace key government reports.

## 1.5 AIDS TO STATUTORY INTERPRETATION

Though it may be possible to find a particular piece of UK legislation, once found it is vital that legislation is interpreted correctly before any further action is taken. Despite a long history of legislation and the best efforts of the Parliamentary Counsel there can, and do, surface problems of statutory interpretation. Thus it is that over the years the courts have established principles and rules to assist them in the interpretation of statutory provisions. Though these rules and principles may be applied by the courts they are of central importance to the professional role of the Environmental Health Officer.

1.5.1 THE LITERAL RULE

This rule of interpretation simply requires that where the wording of statute is clear and unambiguous then those words should be given their literal meaning. The assumption behind this is simple: Parliament says what it means and means what it says. It is not therefore for the courts, nor for others, to hold that a clear and unambiguous expression of Parliamentary will should be ignored. In the *Sussex Peerage Case* (1844) Tindal CJ stated that the:

> ... only rule for the construction of Acts of Parliament is that they should be construed according to the intent of the Parliament

which passed the Act. If the words of the statute are in themselves precise and unambiguous, then no more can be necessary than to expound those words in their natural and ordinary sense. The words alone do, in such cases, best declare the intention of the law-giver.

Such an approach has, however, been criticised as expecting an 'unattainable perfection' in draftsmanship. In many instances the only rule that is necessary is the literal rule. Many statutory provisions are clear and unambiguous and for them to be given force of law does not require 'unattainable perfection'. For example in section 79 of the Environmental Protection Act 1990 one of the categories of statutory nuisance is 'any premises in such a state as to be prejudicial to health or a nuisance'. The term premises is defined in s.79(7) of the Act as including land and any vessel. As the Act refers to 'premises in such a state' on a literal interpretation it is clearly the premises (as a whole) that is to be considered as the statutory nuisance and not the separate items or defects which may be the cause of the problem. Sometimes the provisions contained in a statute are much less clear than this. In these cases it is necessary to consider alternative approaches.

### 1.5.2 THE GOLDEN RULE

This is sometimes called the rule to avoid absurdity. Here, if a statute, on a literal interpretation, would bring about a manifestly absurd result then that interpretation need not be followed and thus that result may be avoided. The practical application is in fact to be found when on a literal interpretation a statute may have two interpretations which could give rise to different outcomes. Where this happens a court may choose which of the two to apply. In the case of *R v Allen* (1872), s.57 of the Offences Against the Person Act 1861 provided that 'whosoever being married, shall marry any other person during the life of the former husband or wife shall be guilty of bigamy'. It was possible here for the word marry to have two meanings. The first, meaning contracting a valid marriage; the second, meaning going through a service of marriage. The court held that the second interpretation should be adopted to avoid the absurd result which must occur if the first were to be applied, i.e. that no one could ever be guilty of an offence.

In *Swansea City Council v Jenkins and Others* (1994), Neath Justices initially upheld a complaint preferred by Mr Jenkins against the council, in that a notice served under s.59(1) of the Building Act 1984 upon him in respect of the cost of works to a private sewer was defective in that it had been served upon premises served by the private sewer only above the breach in that sewer. The notice was sent to all owners of properties served by the sewer above the breach stating that the sewer provided for their property was prejudicial to health and required them to carry out remedial work. The notice was not served upon premises drained by the sewer below the breach, as the council thought that the sewer serving those buildings was not defective.

Section 59 of the 1984 Act provides: '(1) If it appears to a local authority that in a case of a building ... a cesspool or other such work or appliance ... provided for the Building is in such a condition as to be prejudicial to health or a nuisance ... they shall by notice require the owner of the building to make satisfactory provision for the drainage of the building, or ... do such work as may be necessary ...'

The High Court allowed the appeal by Swansea City Council against the decision of Neath Justices. Mr Justice Macpherson said that he disagreed with the conclusion reached that, on a strict interpretation of s.59, the notice should be served upon the owners of the building served by the sewer which was in such a condition as to be prejudicial to health or a nuisance. Sewerage was provided for and served all the buildings linked to the private sewer. If a reasonable jury were asked what this sewer provided and for which buildings, they would say all the houses from which the drainage flowed into it. That was also the logical and rational reading of section 59. To hold otherwise would do violence to the language of the section. Therefore where a private sewer was in such a condition as to merit action under the provisions of the statute, the local authority was entitled to serve a notice to repair upon the owners of the property above the defect and not on all those who were served by the drainage system.

### 1.5.3 THE MISCHIEF RULE OR THE RULE IN HEYDON'S CASE

Under this rule a court will seek to establish the mischief the Act was passed to remedy and then construe the Act so as to deal with that mischief.

In *Heydon's Case* (1584) the Exchequer Chamber laid down rules as follows:

> That for the sure and true interpretation of all statutes in general (be they penal or beneficial, restrictive or enlarging of the common law) four things are to be discerned and considered:
>
> 1. what was the common law before the Act was passed
> 2. what was the mischief and defect for which the common law did not provide
> 3. what remedy the Parliament hath resolved and appointed to cure the disease of the commonwealth
> 4. the true reason of the remedy

And then the office of all the judges is always to make sure construction as shall suppress the mischief and advance the remedy and to suppress subtle inventions and evasions for the continue of the mischief, and *pro privato commodo*, and to add force and life to the cure and remedy according to the true intent of the makers of the Act *pro bono publico.'*

In *Gardiner v Sevenoaks RDC* (1950) the authority, acting under the Celluloid and Cinematograph Film Act 1922, served a notice on the occupier of a cave used for the storage of film. The notice required the recipient to comply with certain safety requirements and described the cave as 'premises'. The recipient of the notice, Gardiner, appealed against the notice on the grounds that the cave could not come within the term premises as used in the Act. The court found that the Act was a safety Act designed to protect third parties and those working where the film was stored. Therefore under the mischief rule the cave was 'premises' for the purposes of the Act.

In the area of health and safety at work the Health and Safety at Work etc. Act 1974 has as its primary purpose the provision of a comprehensive system for dealing with the health and safety of people at work and members of the general public, the mischief here arguably being the lack of protection afforded to workers and the general public. Section 3 of the Health and Safety at Work etc. Act 1974 contains provisions relating to the general duties of employers and the self-employed to persons other than their employees when conducting their undertakings. In *R v Mara* [1987], Mara was the owner of a shop cleaning company. He had a contract with a branch of International Stores to clean their shop and loading bay. For this work electric scrubbing machines were used. It was intended that Mara's employees were to be out of the store by 9.00 am. Unfortunately this proved not to be possible in the loading bay due to interruptions by deliveries. A new arrangement therefore was made, whereby Mara's scrubbing machine was to be made available for use by the employees of International Stores, who were to clean the bay when it was convenient. On one such occasion one of International's employees suffered an electric shock. Investigation revealed that the electrical cable to the machine was damaged and had been taped up in a number of places. In one damaged area bare wire was in fact showing. Mara admitted knowing of at least two places where tape was used on the cable and that large quantities of water were used in the cleaning process. However as the accident happened on a Saturday, when none of Mara's employees went to the store, he argued that on that day he was not conducting his business and therefore s.3(1) of the Health and Safety at Work etc. Act 1974 did not apply. The court found that for the purpose of s.3(1) the conduct of a company's undertaking was not confined to the hours when the company's employees were present. It found that it was not permissible to treat the section as being applicable only when an undertaking was in the process of being actively carried on. Were the court to have accepted Mara's argument they would surely have failed to address the mischief at which the Act was aimed.

### 1.5.4 THE EJUSDEM GENERIS RULE

There exists a problem in the use of lists in statutory provisions. If a list of any kind is used in a statute the question arises: is this an exclusive or merely indicative list of those matters covered by the provision? Exclusive lists have obvious advantages. If the list is exclusive then, as a matter of interpretation, if the thing in question is not on the list it is not regulated by the statutory provision. However as a practical proposition such lists are often impossible. Non-exclusive lists have the obvious advantage of being easily made but the disadvantage of imprecision. Thus it is that the *ejusdem generis* rule can be applied giving clarity to otherwise obscure provisions. Translated this expression becomes 'of the same kind'. What it means is that in cases where general words follow specific words then those general words are to be construed *ejusdem generis* (of the same kind) as the specific. Thus a reference in a statute to 'lions, tigers, bears and other animals' would include creatures *ejusdem generis* with lions tigers and bears, that is other 'tooth and claw' animals, but would not include goldfish, hamsters etc. In *Powell v Kempton Park Racecourse Co. Ltd.* [1899] the House of Lords had to consider s.1 of the Betting Act 1853. This section prohibited the keeping of a 'house, office, room or other place' for betting with persons resorting thereto. The question before the court became, was Tattersall's ring on a racecourse such a place. The Lords said it was not *ejusdem generis* and the words 'or other place' meant a place similar to a house, office or room. This rule does, however, have its limits. In *Allen v Emmerson* [1944], s.33 of the Barrow-in-Furness Corporation Act 1872 stated that 'no theatre or other place of public entertainment' could operate without a licence. The court held that a funfair could not be read *ejusdem generis* with the word theatre. The court noted that no case had been cited in which a genus had been held to be constituted, not only by the enumeration of a number of classes followed by the words 'or other' but by the mention of a single class (in this case a theatre) followed by those words.

### 1.5.5 NOSCITUR A SOCIIS

Associated with the *ejusdem generis* rule is a similar approach to interpretation, that of *noscitur a sociis*. Here the meaning of a word is to be gathered from its context or in the more expressive language of Stamp J in

*Bourne v Norwich Crematorium Ltd.* [1967] words 'derive colour from those which surround them'. In *Pengelley v Bell Punch Co. Ltd.* [1964] the court was obliged to construe s.28(1) of the Factories Act 1961. This section states that:

> All floors, steps, stairs, passages and gangways shall be of sound construction and properly maintained and shall, so far as is reasonably practicable, be kept free from any obstruction and from any substance likely to cause a person to slip.

The word floor is not defined in the Act. The court, using *noscitur a sociis*, found that the word 'floors' used here, when looked at in context, did not apply to a part of a factory used for storage rather than passage.

### 1.5.6 THE EXCLUSIONARY RULE

Normally a court under the 'exclusionary rule' may not refer to parliamentary material as an aid to statutory construction, i.e. how to interpret the Act. However in the case of *Pepper v Hart* [1993] this principle was relaxed. The case itself concerned the question of whether the payment of reduced fees in respect of the sons of teachers at an independent school amounted to a taxable benefit under the Finance Act 1976. This raised the question whether reference could be made to the parliamentary proceedings which led to the passing of those provisions, when construing them.

The court held that the previous position of the House would be relaxed and that such a reference could be made in certain circumstances. Lord Browne-Wilkinson stated that:

> Under present law there is a general rule that reference to parliamentary material as an aid to statutory construction is not permissible (the exclusionary rule) . . . This rule did not always apply but was judge made . . . The courts can now look at white papers and official reports for the purpose of finding the 'mischief' sought to be corrected, although not at draft clauses or proposals for the remedying of such mischief. Given the purposive approach to construction now adopted by the courts in order to give effect to the true intentions of the legislature, the fine distinctions between looking for the mischief and looking for the intention in using words to provide the remedy are technical and inappropriate. Clear and unambiguous statements made by ministers in Parliament are as much background to the enactment of legislation as white papers and parliamentary reports. The decision in *Pickstone v Freemans plc* [1989] which authorises the court to look at ministerial statements made in introducing regulations which could not be amended by

Parliament is logically indistinguishable from such statements made in introducing a statutory provision which, though capable of amendment, was not, in fact amended ... I therefore reach the conclusion ... that the exclusionary rule should be relaxed so as to permit reference to parliamentary materials where:

(a) legislation is ambiguous or obscure, or leads to an absurdity
(b) the material relied on consists of one or more statements by a minister or other promoter of the Bill together if necessary with such other parliamentary material as is necessary to understand such statements and their effects.
(c) the statements relied on are clear.

The purposive approach to construction here refers to the courts' willingness to look at the objectives and purpose of the provisions in question as well as the instrument as a whole (*Litster v Forth Dry Dock & Engineering Co. Ltd.* (1989)).

On 20 December 1994 Lord Taylor handed down a Practice Direction dealing with cases where there were intended references to Hansard (daily transcript of Parliament) in civil and criminal proceedings. This requires that any party to proceedings that is intending to refer to any extract from Hansard must, unless the judge otherwise directed, serve upon all other parties and the court copies of any such extract. As an aid to interpretation of a statutory provision this is a very useful tool for the Environmental Health Officer.

### 1.5.7 OTHER AIDS

In addition to the rules indicated there are a number of other aids to statutory interpretation which can assist in establishing the compass of a statutory provision.

### 1.5.8 INTERPRETATION ACTS

The Interpretation Act 1978 aims to give certain standard definitions for common provisions found in statutes. Section 6 of the Act, for example, provides that:

In any Act, unless the contrary intention appears–

(a) words imparting the masculine gender include the feminine;
(b) words imparting the feminine gender include the masculine;
(c) words in the singular include the plural and words in the plural include the singular.

The courts have considered the application of the Interpretation Act to clarify statutory provisions. Under the Water Industry Act 1991, ss. 158 and 159 a sewerage undertaker may construct public sewers through private land after giving reasonable notice to the owner occupier. In *Hutton v Esher UDC* [1973] the local authority proposed to construct a sewer under similar provisions found in the earlier Public Health Act 1936. The proposed line of the sewer would have taken it through Hutton's bungalow. For this to occur the bungalow would have to be demolished, though it could be rebuilt. The Public Health Act 1936 gave the authority the power to build a sewer 'in, on, or over any land not forming part of a street'. The plaintiff contended that the expression 'land' did not extend to buildings and therefore there could be no power to demolish his house. Section 3 of the Interpretation Act 1889 (now to be found in s.5 and Schedule 1 to the Interpretation Act 1978) provided that, unless a contrary intention appears, the expression 'land' includes buildings. The Court of Appeal held the Interpretation Act to be applicable and therefore the local authority indeed had the power to demolish the plaintiff's house.

### 1.6 JUDICIAL PRECEDENT

In essence this principle means that like cases will be treated alike, and established rules of law and practice will be followed. Derived from Custom, the doctrine of *stare decisis* is essential in order that the law may have some degree of predictability and certainty. Were there to be no predictability, no certainty, then the only way that two parties could possibly know the likely outcome in a legal dispute would be to go before the courts and have the matter decided. Theoretically this may appear perfectly sound, however in practice such a system would quickly prove to be inoperative in terms of both cost and time. The predictability associated with a system of precedent may be thought of as one of the oils which lubricate the machine of justice; a machine which, though it grinds slowly, might prove to move at a more glacial pace were precedent not to be present. Were it not to exist then lawyers could not advise their clients and Environmental Health Officers could not take many of the day to day decisions that they do. The certainty which comes with such a system is also however its drawback. It can make the law inflexible and appear to be out of time with the society it is meant to serve. This is not however inevitable and an understanding of the method of operation of the system of precedent can reveal how it is able to accommodate the apparently opposite concepts of certainty and flexibility into a workable system of law. Precedent is sometimes referred to as the law made by judges or the common law. Strictly precedent is not common law, though the rules of the common law have been incorporated into judicial decisions and thus become part of the system of judicial precedent to the point that it is hard for the layman to see the difference. The system of precedent as operated today requires two elements to be in place before it can relied upon; firstly a good system of high quality law reports and secondly a clear and established hierarchy of the courts.

### 1.6.1 LAW REPORTS

Firstly it important to understand that these are not a verbatim transcript of everything that was said in a case. The evidence of witnesses, the expertise of cross examination, the interruptions from the well of the court, none of these are to be found here. What is recorded and reported is much narrower than that. Indeed only a very small proportion of all cases are reported in full. Of those that do reach the senior courts only something less than 10% are reported.

In addition there is no governmental or, in that sense, official control over the reporting of cases. However since the middle of the last century law reporting has been under the control of the Incorporated Council of Law Reporting. This is a committee of the Bar Council, Law Society and the Inns of Court. Reports subject to supervision here are scrutinised by the judge who heard the case before publication. Generally speaking a case will be the subject of reporting if it involves or raises a significant point of law. Cases which therefore extend, restrict, or confirm an existing principle of law or create a new principle will generally find themselves reported.

Those who publish reports privately, for example the *All England Reports*, decide themselves on what is and is not to be reported. Guidance given by the All England Law Reports states that a case should:

- Make new law by dealing with a novel situation or extending the application of existing principles
- Give a modern judicial restatement of existing principles
- Clarify conflicting decisions of lower courts
- Interpret legislation likely to have wide application
- Interpret a commonly found clause
- Clarify an important point of practice or procedure.

### 1.6.2 CITATION OF LAW REPORTS

Series of law reports are normally referred to by abbreviations of their title. The full reference to the report of a case in a series of law reports is called a citation. Before a report can be found the citation must be understood.

Cases are usually referred to (cited) in this way: **National Coal Board v Thorne [1976] 1 WLR 543.**

- **National Coal Board v Thorne.** This is the names of those parties involved in the case. For civil matters these are the plaintiff versus (v) the defendant. In a criminal case the notation will be R v Smith, indicating that the case was taken by the Crown (R for Rex or Regina). It is important to note that sometimes a case may be reported by more than one name; thus N.C.B. v Thorne [1976] 1 WLR 543 is sometimes referenced under another name *(Sub nom)* N.C.B. v Neath BC [1976] 2 All ER 478. The name of the plaintiff should be the first name and the defendant the second.
- **[1976].** There will then be the year that the case was reported. This is given in square brackets and is an essential part of the reference. If the date is given in round brackets this indicates the date of hearing, which is not necessarily the same as the date when it was reported; the date in round brackets is not an essential part of the citation of the case reference. Some older reports did not use the year as part of the reference and preferred instead to number the volume of the reports sequentially.
- **1.** This is the volume number of the report series.
- **WLR.** This is the abbreviated reference for the series of law reports which contain the report. There are a number of series of reports, some of which are reproduced later.
- **543.** Finally there is presented the page number where the report is to be found.

  Thus in the above example the case of *National Coal Board v Thorne* is to be found in the first volume of the *Weekly Law Reports* for the year 1976 at page 543; and in the second volume of the *All England Reports* for the year 1976 at page 478.

It is possible to indicate the relevant passage of a judgment in a case, for example Lord Justice Thesiger's famous speech on the significance of locality to the assessment of nuisance runs as follows:

> What would be a nuisance in Belgrave Square would not necessarily be so in Bermondsey: and where locality is devoted to a particular trade or manufacture carried on by traders or manufacturers in a particular and established manner not constituting a public nuisance, Judges and juries would be justified in finding, and may be trusted to find, that the trade or manufacture so carried on in that locality is not a private or actionable wrong.

This is to be found at page 865 in the report. The full reference for this is thus *Sturgess v Bridgeman* (1879) 11 ChD 852, 865. It can be seen that the last series of numbers indicates where the words are to be found in a report which starts on page 852. The use of round brackets indicates here that the date is not an essential part of the reference as the report gives a number (11) to the volume of the report.

Sometimes it may be possible to be even more accurate in referencing as some of the reports will have the individual paragraphs labelled by letter. Thus in *N.C.B. v Thorne* [1976] Watkins J, at page 482c, establishes the stated interpretation of nuisance against a self-confessed attraction to the idea that '... whatever is complained about must in some way be directed to the question of the health of the person who claims to be or has been affected by the nuisance.' The letter 'c' after the page number here indicates where precisely the paragraph containing the quote is to be found on the page.

Some of the most common law report abbreviations likely to be encountered by the Environmental Health Officer are reproduced below.

| | |
|---|---|
| AC | Law Reports – Appeal Cases |
| All ER | All England Law Reports |
| CMLR | Common Market Law Reports |
| ChD | Law Reports – Chancery Division |
| CrAppR | Criminal Appeal Reports |
| ECR | European Court Reports |
| ER | English Reports |
| EnvLR | Environmental Law Reports |
| IRLR | Industrial Relations Law Reports |
| ICR | Industrial Cases Reports |
| JPR | Justice of the Peace Reports |
| KIR | Knights Industrial Reports |
| Knight's Ind & Com | Knight's Industrial and Commercial Reports |
| QBD/KBD | Law Reports – Queens/Kings Bench Division |
| TLR | Times Law Reports |
| WLR | Weekly Law Reports |

## 1.7 THE LAYOUT OF A LAW REPORT

Though different series of reports will be presented in differing ways, there are certain elements of a report which are repeated. These are:

1. The names of the parties involved in the case. Plaintiff first, defendant second. For cases on appeal these carry the same title as that which applied when the case was first heard. This means that the plaintiff is still shown first whether they be appellant or respondent in the Court of Appeal or the House of Lords. In the case of the House of Lords this has only been the case since 1974 (Practice Direction [1974] 1 WLR 305).

2. The court in which the case was heard.
3. Who the judge(s) were.
4. The dates of hearing and, if these are not the same, the date of the judgment. If the words *Cur adv volt* appear this is an indication that the judgment was reserved i.e. given at a later date.
5. The report will also use keywords added by the law reporter. These are the so called 'catchwords' which summarise the legal issues involved.
6. There is a Headnote, which will summarise the facts of the case leading to the action.
7. There is then to be found the decision of the court. All cases referred to in the judgment are then listed. These cases may or may not have been relied upon to arrive at the judgment in the case being reported. Also recorded are cases which would have been referred to in the arguments submitted by the lawyers acting for the parties.
8. There will then follow the judgment. It is here that the judge or judges will give, with reasons, their findings.
9. The final part will contain details of the order made by the court and the names of the solicitors acting for the parties.
10. Last of all will come the name of the barrister reporting the case.

## 1.8 FINDING A LAW REPORT

There are two basic situations when a report is sought. The first is when the name of the case is known but not the details which it contains. The second situation is when research is undertaken in a subject area and relevant cases on that subject are sought. Both require access to a specialist law library.

### 1.8.1 WHEN THE NAMES ARE KNOWN

Here it is best to use the *Current Law Case Citator*. This includes an alphabetical listing of all cases reported since 1947. A case is listed with a full report reference which will allow the full text to be obtained. If the case pre-dates 1947 then 'The Digest' should be used. This contains summaries of cases going back many hundreds of years.

### 1.8.2 WHEN A SUBJECT IS BEING RESEARCHED

This is the harder of the two methods of research. Again the Digest may be of use. *Halsbury's Laws of England* is also very useful. Halsbury's is a multi-volume set in which subjects are arranged alphabetically. The text is referenced and includes principal cases. There are also an increasing number of CD-ROM packages coming on line and these will help to make searching much easier in the future.

## 1.9 THE HIERARCHY OF THE COURTS

The system of precedent, in addition to requiring a system of law reporting, requires that there be an established hierarchical system of courts. Within the English court system there is a division between courts of criminal jurisdiction and those of civil jurisdiction.

### 1.9.1 THE CIVIL COURTS

#### 1.9.1.1 The County Court

This was established in 1846 to deal with small claims arising out of civil disputes. The court is now regulated by the County Courts Act 1984, the Courts and Legal Services Act 1990 and the High and County Courts Jurisdiction Order 1991. It deals with matters relating to contractual disputes, debt, hearings under the Children Act and undefended divorce proceedings. It is presided over by a circuit judge who usually sits alone. Appeal lies from this court to the Court of Appeal (Civil Division). The court has jurisdiction to hear claims for damages in respect of personal injury or death under £50 000 and other actions, except actions for defamation, where the sum involved is less than £25 000. Above this court stands the High Court.

#### 1.9.1.2 The High Court

The High Court is subdivided into three divisions: Queens Bench Division, Family Division and Chancery Division.

#### 1.9.1.3 Queens Bench Division

The Queens Bench Division (QBD) is competent to hear all civil claims. It is the largest of the three divisions having some 50-plus judges and is presided over by the Lord Chief Justice. The QBD will deal with all civil matters not put before the other two divisions. The QBD is thus the division which will hear matters of contract or tort.

#### 1.9.1.4 Family Division

This division, which is presided over by the President of the Division, deals with family law matters, for example wardship and guardianship are dealt with here. It also hears matters related to divorce i.e. division of property and access to children.

### 1.9.1.5 Chancery Division

The Chancery Division is the smallest of the divisions and deals with wills and probate, trusts, tax, mortgages etc. There are 13 judges who sit here and it is presided over by the Lord Chancellor, though in reality it is the Vice Chancellor who is appointed to deal with organisational and administrative matters.

All three divisions can hear cases at first instance and on appeal. When hearing an appeal, the High Court is called the Divisional Court. The Divisional Court of the Queens Bench Division hears appeals on points of law from the Magistrates and Crown Courts. The Divisional Court of the Chancery Division hears appeals in bankruptcy from the County Court. The Divisional Court of the Family Division hears appeals on family matters dealt with by the magistrates courts.

### 1.9.1.6 The Court of Appeal (Civil Division)

The court is presided over by the President of the Civil Division who is the Master of the Rolls. There are a maximum of 28 judges, Lords Justices of Appeal, that sit in this court. The Civil Division of the Court of Appeal can hear appeals from any division of the High Court and from the County Court. It will normally sit as a court of three judges but can sit as a court of five or seven. The more important sittings of the court will be headed by the Master of the Rolls. On hearing an appeal the court may uphold or reverse a decision of a lower court or it may substitute a new judgment of its own.

### 1.9.1.7 House of Lords

This is both the second chamber of Parliament and the highest court in the land. As a court of law, as distinct from the second chamber of the legislature, it consists of the Lord Chancellor and the Lords of Appeal in Ordinary. There are a maximum of 11 Lords of Appeal in Ordinary (Administration of Justice Act 1968, s.1).

### 1.9.1.8 Civil Appeals

The House of Lords will hear appeals from both the Criminal and Civil Divisions of the Court of Appeal.

The House of Lords will hear appeals from the Court of Appeal and the Court of Session in Scotland. There also exists the 'leap frog' procedure whereby an appeal is permissible from the High Court to the House of Lords.

### 1.9.1.9 Criminal Appeals

The court will hear appeals from the Criminal Division of the Court of Appeal and the Divisional Court.

1.9.2 THE CRIMINAL COURTS

### 1.9.2.1 The Magistrates Court

This is primarily but not exclusively a court of criminal jurisdiction. It retains some responsibility in family and licensing matters. All criminal cases start in the magistrates courts and it is in the magistrates courts that most criminal cases are dealt with. The Magistrates Court has criminal jurisdiction to hear summary offences and either way offences.

These offences are defined by reference to s.64 of the Criminal Law Act 1977 as amended by the Interpretation Act 1978, Schedule 1, in the following way:

(a) an indictable offence is an offence which if committed by an adult is triable on indictment whether it is exclusively so or triable either way;
(b) a summary offence is an offence which if committed by an adult is triable only summarily; and
(c) an either way offence is an offence which if committed by an adult is triable on indictment or summarily.

If the offence is purely a summary offence s.2 of the Magistrates Court Act 1980 requires that the offence be tried in the county or London Commission area within which it was committed. Indictable offences, which are offences triable on indictment (whether exclusively so triable or triable either way), can be tried in any Magistrates Court providing the jurisdiction of the criminal law applies to those offences (Magistrates Court Act 1980, s.2). A summary offence is an offence mentioned in column 1 of Schedule 1 to the Criminal Law Act 1977 (as amended). An offence triable either way is an offence defined by s.17 of the Magistrates Court Act 1980 and set out in Schedule 1 to that Act or an offence made so triable by the legislation which creates the offence.

For any Act encountered by an Environmental Health Officer the status of an offence will be explicitly stated. For example s.33 of the Health and Safety at Work etc. Act 1974 makes it clear that preventing or attempting to prevent any person from appearing before or answering any question put by an inspector under the provisions of the Act is a summary offence carrying a maximum penalty of £5000. Whereas the more serious either way offences, for example acquiring or attempting to acquire an explosive article or substance in contravention of a relevant statutory provision, are to be found in section 33(4) and carry with them the sanction of an unlimited fine, or two years imprisonment, or both.

### 1.9.2.2 The Crown Court

This court was established under the Courts Act 1971, and replaced the former Courts of Assize and Quarter Session. Cases are heard by Judges, Recorders or High Court Judges depending on the seriousness of the case. The Crown Court can hear appeals from the Magistrates Court and will also deal with committals for sentence from that court. Such committals arise when, after dealing with a trial in the magistrates court, the magistrates decide that their sentencing powers are inadequate to deal with the offender. When this occurs the case is then remitted to the Crown Court for sentence. The main body of the work undertaken in this court will deal with trials of indictable offences, as previously defined.

### 1.9.2.3 The Divisional Court

The Divisional Court of the QBD, as already indicated, will hear appeals on points of law from Magistrates and Crown Courts.

### 1.9.2.4 The Court of Appeal (Criminal Division)

The Criminal Division of the Court of Appeal hears appeals from the Crown Court against conviction or sentence. This is regulated by Part I of the Criminal Appeal Act 1968. The Attorney General may also refer matters to the court under s.36 of the Criminal Justice Act 1972. The Home Secretary may also refer a case to the court under the provisions of s.17 of the Criminal Appeal Act 1968.

Section 36 of the Criminal Justice Act 1988 permits the Attorney General to refer a case to the Court of Appeal if it appears that the sentence imposed has been too lenient.

### 1.9.3 THE COURT OF JUSTICE OF THE EUROPEAN COMMUNITIES

The European Court of Justice (ECJ) sits in Luxembourg and is competent to rule on the interpretation and application of Community laws. It has 15 judges including one from each member country. Though not strictly part of the hierarchy of the English court system it is important, not least because the judgments of the Court are binding in the UK.

The jurisdiction of the Court is conferred by the Treaties and can be classified under three main headings.

### (1) Action against a Member State or Community Institutions

Here the Court will hear referrals from the Commission under Article 169 of the Treaty of Rome or from one of the member states under Article 170. Under Article 169, if the Commission considers that a member state has failed to discharge its obligation under Community law, it must give an opinion on the matter. If the offending state still does not comply with the opinion, within a specified period, the Commission may put the matter before the ECJ. Under Article 170 similar proceedings may be brought by a member state against another member state. A member state is obliged to comply with the declaration of the Court.

### (2) Opinions

Here the Court examines the legislation and the decisions of the Council and the Commission and expresses an opinion.

### (3) Applications for Preliminary Rulings

Article 177 of the EEC Treaty provides the European Court of Justice with jurisdiction to give preliminary rulings concerning:

(a) the interpretation of the treaties;
(b) the validity and interpretation of acts of the institutions of the community.

Where such a question is raised before any court of a member state, that court may, if it considers that a decision on the question is necessary to enable it to give judgment, request the Court of Justice to give a ruling thereon.

This ability to refer to the ECJ exists to ensure a uniformity of interpretation across the European Union. Therefore when cases come before an English court which involves a clarification of EU law the matter may then be referred to the ECJ and a ruling given. This will of course be applicable to the case in question. However, thereafter that ruling will not only guide English courts but all courts of member states when considering that same issue. Similarly an English court will be influenced by the outcome of a referral by a court from any other member state. The ECJ is clearly not therefore an appellate court so far as the English hierarchy of courts is concerned, however its importance is clear. It is the only court which can give an authoritative interpretation of treaty provisions of the EU and, once given , such rulings will influence domestic courts and tribunals.

### 1.9.3.1 The Court of First Instance

Under Articles 4, 11, and 26 of the Single European Act provision was made for the establishment of a Court of First Instance, which is attached to the European Court of Justice.

The Court of First Instance (CFI) was established in 1988 by Article 168A and Council Decision 88/591 to relieve the ECJ of the need to deal with the less important cases coming before it. The CFI has a limited jurisdiction, being able to deal only with competition matters and staff disputes.

### 1.9.3.2 The European Court of Justice and Precedent

Based on a different legal system to our own the Court clearly has the ability and authority to depart from its previous decisions. In practice it does not often do so. However because of the non-binding nature of the decisions a member state is not barred from requesting a ruling under Article 177 merely because the matter has previously been decided by the European Court of Justice.

### 1.10 THE APPLICATION OF JUDICIAL PRECEDENT

In *Ashville Investments Ltd. v Elmer Contractors Ltd.* [1988] the Court of Appeal considered the question of precedent. Lord Justice May observed that:

> it is necessary carefully to consider the role of precedent and the doctrine of *stare decisis* . . . In my opinion the doctrine of precedent only involves this; that when a case has been decided in a court it is only the legal principle or principles on which that court has decided that bind courts of concurrent or lower jurisdiction and require them to follow and adopt them when they are relevant to the decision in later cases before those courts. The *ratio decidendi* of a prior case, the reason why it was decided as it was, is in my view only to be understood in this somewhat limited sense.

Binding precedent is rendered ineffective if it is overruled by a superior court or by a statute. However where the precedent has binding force then it must be followed unless the case in hand can be distinguished on its facts. For example in *R v Parole Board, ex parte Wilson* [1992] a man contended that before a review he was entitled to see the reports which were to be put before a Parole Board. In an earlier case, *Payne v Lord Harris of Greenwich* [1981] the Court of Appeal had decided that a man serving a mandatory life sentence was not so entitled. In the instant case the Court of Appeal held itself not bound by the *Payne* case as that decision related to a mandatory life sentence here there was a discretionary life sentence and it was therefore distinguishable.

There are two kinds of precedent: binding and persuasive precedent.

### 1.10.1 BINDING PRECEDENT

This means that a court when confronting a case on a matter which has been previously considered by another court must not only refer to that or those previous decisions but must also apply the law in those decisions to the case before them. The question then arises: how can a judge know which cases bind and which do not?

Whether a particular statement by a judge is binding or not will depend on whether it is termed the *ratio decidendi* (reason for the decision) or whether it was merely something said *obiter dictum* (by the way). It is the *ratio decidendi* which is the binding element of a previous case. The relative status of the courts involved is also vital to the issue.

### 1.10.1.1 Ratio Decidendi

A judgment will contain the following:

(a) A statement of the facts and an indication of which of them are material.
(b) A statement of the legal principles involved which apply to issues raised by the material facts. This part of the judgment constitutes the *ratio decidendi* and can have binding effect.
(c) The actual judgment, decree, or order. This is binding only on the parties.

It is often difficult to ascertain the *ratio decidendi* of a case and this is one of the skills of the legal expert.

### 1.10.1.2 Obiter Dicta

These are legal reasons or pronouncements which are said 'by the way'. Examination of these elements of the judgment will reveal that they are not necessary to the decision in the case and therefore not part of the *ratio decidendi* and not binding. The case of *Quick v TaffEly DC* [1986] concerned an alleged breach of section 32 of the Housing Act 1961 (now s.11 of the Landlord and Tenant Act 1985 as amended by Housing Act 1988, s.116). This section provides that, in the case of a letting of a dwelling house for a term of less than seven years, the landlord impliedly covenants to keep in repair the structure and exterior of the house, including drains, appliances for the supply of water, gas, electricity, sanitation and space for heating water. The house in question suffered from severe condensation, so much so that Dillon LJ concluded at page 815: '… that by modern standards the house was in winter, when of course the condensation was worst, virtually unfit for human habitation.'

He went on:

> When I read the papers in this case I was surprised to find that the plaintiff had not based his claim on an allegation that at all material times the house let to him by the defendant council had not been fit. The uncontradicted evidence accepted by the trial Judge showed that furniture, furnishings and clothes had rotted because of damp and the sitting room could not be used because of the smell of damp.

These remarks are obiter since they do not relate to the alleged breach of the repairing obligation. Yet even as *obiter dicta* their value is clear. They give an insight into those matters the judge, and therefore the Environmental Health Officer, might have regard to in assessing matters relating to fitness should they arise in a later case. This is therefore highly persuasive. Persuasive precedents are statements which a later court will respect but need not follow. These will include the following:

(a) *obiter dicta*
(b) *ratio decidendi* of inferior courts
(c) *ratio decidendi* of Scottish, Commonwealth or foreign courts
(d) statements of the Judicial Committee of the Privy Council.

### 1.10.2 THE DOCTRINE OF PRECEDENT AND THE COURTS

The doctrine of precedent depends on the fact that each court in our system stands in a defined relative position to other courts and those positions affect the status of the decision, i.e. whether or not it is binding or persuasive.

### 1.10.3 HIERARCHY OF PRECEDENT

#### 1.10.3.1 The European Court

As already indicated the decisions of the ECJ bind all UK courts but it is not obliged to follow its own previous decisions.

### 1.10.3.2 The House of Lords

Decisions of the House of Lords bind all English courts but since a Practice Statement in 1966 the House of Lords does not have to follow its own decisions.

In that Practice Statement the House of Lords noted that:

> Their Lordships regard the use of precedent an indispensable foundation upon which to decide what is the law and its application to individual cases. It provides at least some degree of certainty upon which individuals can rely in the conduct of their affairs, as well as a basis for orderly development of legal rules.

> Their Lordships nevertheless recognise that too rigid adherence to precedent may lead to injustice in a particular case and also unduly restrict the proper development of the law. They propose therefore to modify their present practice and, while treating former decisions of the House as normally binding, to depart from a previous decision when it seems right to do so. In this connection they will bear in mind the danger of disturbing retrospectively the basis on which contracts, settlements or property and fiscal arrangements have been entered into and also the especial need for certainty as to the criminal law.

This announcement was not intended to affect the use of precedent elsewhere than in the House of Lords.

### 1.10.3.3 The Court of Appeal

The Civil Division of the Court of Appeal will consider itself bound by its own previous decisions unless certain conditions apply. In *Young v Bristol Aeroplane Co. Ltd.* [1944] there was an appeal by the plaintiff against a decision that he was not entitled to damages for breach of statutory duty. As part of this it was argued that the Court of Appeal was not bound by its own previous decisions. The court held that subject to three exceptions it was bound. The Master of the Rolls Lord Greene stated that:

> On a careful examination of the whole matter we have come to the clear conclusion that this court is bound to follow previous decisions of its own as well as those of courts of co-ordinate jurisdiction. The only exceptions to this rule . . . we here summarise:
>
> - the court is entitled and bound to decide which of two conflicting decisions of its own it will follow

- the court is bound to refuse to follow a decision of its own which, though not expressly overruled, cannot in its opinion stand with a decision of the House of Lords
- the court is not bound to follow a decision of its own if it is satisfied that the decision was given *per incuriam.*

The court is justified in not following one of its own previous decisions where it has given its decision in ignorance or forgetfulness of some inconsistent statutory provision or some authority binding on it or if the decision involved some manifest slip or error (*Rickards v Rickards* [1989]).

So far as the Criminal Division is concerned it was stated in *R v Gould* [1968] by Lord Diplock that the Court of Appeal, in its criminal jurisdiction, does not apply the doctrine of *stare decisis* as rigidly as in its civil jurisdiction. If it is of the opinion that the law has been misapplied or misunderstood it will depart from a previous decision; this is irrespective of whether or not the case comes within any of the exceptions contained in *Young v Bristol Aeroplane*. In *R v Spencer* [1985] Lord Justice May stated that it was:

> difficult to see why there should in general be any difference in the application of the principle of *stare decisis* between the Civil and Criminal Divisions of [the Court of Appeal], save that we must remember that in the latter we may be dealing with the liberty of the subject and if a departure is necessary in the interests of justice to an appellant then this court should not shrink from so acting.

### 1.10.3.4 The High Court

The Divisional Court is bound by the decisions of superior courts and bound by its own previous decisions subject to the same rules as are applied in the Court of Appeal. Decisions taken by the court sitting at first instance are not binding on other High Court judges and are of persuasive authority only.

### 1.10.3.5 The Crown Court

This must follow earlier decisions of the House of Lords and the Court of Appeal.

### 1.10.3.6 The Inferior Courts

Magistrates Courts, County Courts and inferior Tribunals are not bound by their own earlier decisions but are bound by the decisions of the higher courts.

## 1.11 EUROPEAN UNION LAW

The law of the EU may be considered as primary legislation, in the form of Treaties, and secondary legislation in the form of Regulations, Directives and Decisions. The European Court of Justice also, as has been noted, produces case law.

### 1.11.1 COMMUNITY TREATIES

This consists of the treaties which established the three Communities: the Treaty of Paris created the European Coal and Steel Community (ECSC), EURATOM established the Atomic Energy Community and the Treaty of Rome created the European Economic Community. There have been various amending treaties. In 1965 a Treaty Establishing a Single Council and a Single Commission of the European Communities was signed. The Treaty of Rome was amended by the Single European Act (SEA) in 1986 and again by the Treaty of Maastricht (the Treaty on European Union (TEU)).

### 1.11.2 SECONDARY LEGISLATION

Whilst the treaties provide for the overall objectives of the EU and give a constitutional basis for those objectives, the secondary legislation provides the mean for achieving those objectives via detailed provisions aimed at specific areas. Article 189 of the Treaty of Rome provides for different kinds of Act which may be adopted by the Council or the Commission. They are:

- Regulations
- Directives
- Decisions
- Recommendations
- Opinions.

### 1.11.2.1 Regulations

Regulations are legislative in effect and lay down rules which are binding on member states. Under Article 189 of the Treaty, regulations are directly applicable without any Act of implementation by member states.

### 1.11.2.2 Directives

A Directive is binding only as to the result to be achieved. It leaves it to the member state to choose the most appropriate domestic legislative mechanism to give effect to the measure. So, for example, the Noise at Work Regulations 1989, which took effect from 1 January 1990, implemented in the UK the provisions of Directive 86/188/EEC (OJ No. L 137, 24/5/86, p.28). This Directive, which was adopted in May 1986, committed member states to having legislation in place by 1990 to deal with the protection of workers from the risks related to exposure to noise at work.

Unlike Regulations, Directives are not binding generally but bind only those member states to whom they are addressed.

### 1.11.2.3 Decisions

Like Directives these are also not generally binding. They apply only to those to whom they are addressed.

### 1.11.2.4 Recommendations and Opinions

Both Recommendations and Opinions are not binding. Simply they give the opinion of the devising organisation.

### 1.11.3 FINDING EUROPEAN LEGISLATION

Secondary legislation of the EU is printed in the Official Journal of the EU (OJ). The covers of these publications are colour-coded according to language, the English language publications having purple covers.

The Journal is divided into three main parts:

(a) the L series (Legislation) which gives the text of the legislation;
(b) the C series (Communications and Information) which contains official announcements and information on EU activities and draft legislation; and
(c) the Annex which holds the text of debates held in the European Parliament.

References to the Official Journal are as follows:
**OJ   Series Number   Date of Issue   Page Number.**
Thus the Directive on air quality standards for nitrogen dioxide is to be found by using the reference OJ. L 087, 27.03.85 p. 0001.

The full citation of European Union legislation should include the following elements:

(a) the institutional origin of the Act (Commission or Council)
(b) the form of the Act (Regulation, Directive, etc.)
(c) the year it was enacted
(d) the unique legislation number
(e) the treaty under which it is made (EEC, ECSC, EURATOM)
(f) the date the legislation was passed.

Thus the earlier reference becomes Council Directive 85/203/EEC of 7 March 1985 on air quality standards for nitrogen dioxide, OJ. L 087, 27.03.85 p. 0001. Within the UK compliance with Directive 85/203 EEC was accomplished by using a variety of pre-existing and new provisions, most particularly the Air Quality Standards Regulations 1989 made under s.2(2) of the European Communities Act 1972.

For Regulations the citation is slightly different. So it is that for Council Regulation (EEC) No. 594/91 of 4 March 1991 on substances that deplete the ozone layer OJ L 067, 14/3/91, and Council Regulation (EEC) No. 3952/92 of 30 December 1992, amending Regulation 594/91 in order to speed up the phasing out of substances that deplete the ozone layer, OJ L 405, 31/12/92, it is important to note that both have their legislation number first, in contrast with Directives and Decisions which have the year first.

### 1.11.4 INTERPRETATION OF EUROPEAN LEGISLATION

As may be seen in the English legal system the rules of statutory interpretation are well recognised and whilst still permitting discretion, act as a guide to the courts. These rules are, as already discussed, assisted but not superseded by the Interpretation Acts. However when EU legislation is to be interpreted things are more problematic. There are not to be found within the EU legal order any Directives relating to interpretation. Article 164 of the Treaty merely states that the European Court of Justice (ECJ); 'shall ensure that in the interpretation and application of this treaty the law is observed'.

Without greater elaboration such a bland statement would be of limited use. However guidance on the ECJ's approach to interpretation is to be found in Case 283/81 *Srl CILFIT v Italian Ministry of Health* [1982]. In this case the Italian Court of Cassation was in doubt whether, as the highest Italian court, it was obliged to refer a question to the ECJ under EEC Article 177(3) and sought the view of the ECJ.

In reaching its decision the Court gave guidance on the approach generally to be adopted when interpreting Community provisions. It held:

To begin with ... community law is drafted in several languages and that the different language versions are all equally authentic. An interpretation ... thus involves a comparison of different language versions.

It must also be borne in mind ... that community law uses terminology which is peculiar to it. Furthermore it must be recognised that legal concepts do not necessarily have the same meaning in community law and in the various member states.

Finally every provision of community law must be placed in its context and interpreted in the light of the provisions of community law as a whole, regard being had to the objectives thereof and to its state of evolution at the date on which the provision in question is to be applied.

## 1.11.5 EUROPEAN UNION CASE LAW

The method of the European Court of Justice (ECJ) is based on a system of law different to that of the UK and reports of cases heard by the ECJ have a different arrangement to that of UK cases. The reports of the ECJ are divided into three parts. These are:

1. The Report for the Hearing
2. The Opinion of the Advocate General
3. The Judgment of the Court.

### 1.11.5.1 The Report for the Hearing

This is a statement made to the Court by a Judge Rapporteur (a reporting judge). It outlines the facts of the case and gives a summary of the legal arguments which are involved.

### 1.11.5.2 The Opinion of the Advocate General

One Advocate General is assigned to each case. The Advocate General holds the position of an 'adviser' to the European Court of Justice.
Article 166 of the Treaty describes his function as:

... acting with complete impartiality and independence, to make, in open court, reasoned submissions on cases brought before the Court of Justice, in order to assist the court in the performance of the task assigned to it in Article 164.

The role of the Advocate General has no direct equivalent in the English legal system. It is the duty of the Advocate General to act with impartiality to assist the court, by making reasoned submissions on cases brought before the Court of Justice. After researching the issue he will give an opinion on it, setting out the facts of the case and all relevant legal provisions. Having regard to previous decisions he will examine the issues and suggest a course of action for the Court. The opinion of the Advocate General is not binding on the Court.

### 1.11.5.3 The Judgment of the Court

This is the part of the report which contains the judgment as delivered by the Court. Unlike an English report there are no dissenting judgments contained in a report of the ECJ; the Court gives only a single judgment.

### 1.11.6 THE CITATION OF ECJ CASE LAW

The decisions of the European Court of Justice are to be found in the official reports of the European Union: the 'European Court Reports'. Citation of European Court cases is different to that of cases in the English courts and takes for example the following format:

*Case C-131/88 EC Commission v Germany* [1991] IECR 825, or
*Case 44/81 Germany v Commission* [1982] ECR 1855.

The citation can be presented as follows:

| Case Registration Number | Name of Parties | Year | ECR | Part Number Only from 1990 | Page Number |
|---|---|---|---|---|---|

*Case Registration Number – Case C-131/88*

Cases which are heard before the European Court of Justice are prefixed by the letter C. Cases coming before the Court of First Instance carry the prefix T. Where a case was heard before the establishment of the Court of First Instance it will carry no prefix.

*Name of Parties – EC Commission v Germany*

As for an English report the names of the parties to the action form part of the reference.

*Year – [1991]*

European Court Reports, which can take up to two years to be published, are not the most up-to-date source of reference. For reports of cases the Common Market Law Reports (CMLR) report European Union cases more rapidly. They are a series of private reports and are issued weekly. Clearly these offer an up-to-date and readily accessible point of reference for Environmental Health professionals.

### I ECR

The European Court Reports [ECR] is divided into two sections. Section I contains those cases which are heard by the European Court of Justice; Section II contains cases put before the Court of First Instance.

### 825

This is the page number at which the case is to be found.

### 1.12 OTHER SOURCES OF LAW

Though the preceding parts of this chapter have identified the sources of 'law proper', nevertheless the Environmental Health Officer must have regard to a number of other sources which, though not in the true sense law, nevertheless are legally significant.

### 1.12.1 CIRCULARS AND CODES OF PRACTICE

In addition to the 'proper' sources of law the departments of government issue numerous circulars and codes of practice.

A government circular is the means by which the government departments most intimately concerned with a piece of legislation seek to give formal and authoritative guidance on that legislation. These documents explain the background to an Act, provide guidance, suggest approaches to interpretation etc. They are not of legal effect, in the sense that to ignore them is not to break the law, but they are strongly persuasive and influential when it comes to the interpretation of a particular statute. Because circulars are the formal means by which government departments give advice as to how to implement and interpret statutes they are uniquely influential in the way a particular piece of legislation will be interpreted and enforced – they are thus of legal consequence.

For example, Department of the Environment Circular 17/96 provides detailed guidance on the interpretation of the fitness standard found in section 604 of the Housing Act 1985. Any Environmental Health Officer wishing to undertake an adequate assessment of fitness must have regard to the Guidance Notes in the Circular. Among other similarly relevant circulars of central importance in understanding the controls applied to housing conditions are:

5/90     Houses in Multiple Occupation
10/90   House Adaptations for Disabled People
12/92   Houses in Multiple Occupation: Guidance to Local Authorities on Standards of Fitness under Section 352 of the Housing Act 1985
12/93   Houses in Multiple Occupation: Guidance to Local Authorities on Managing the Stock in Their Area.

Similarly, though in an entirely different area, the appropriate use of Formal Cautions can only be understood by having regard to the contents of Home Office Circular 18/1994.

Though codes of practice may be encountered by the Environmental Health Officer in a number of areas, food, housing etc., it is arguable that it is in the area of health and safety at work that the code of practice is most commonly encountered. The Robens Committee in its report (Cmnd 5034) highlighted the scope for making explicit reference in statutory regulations to codes of practice as a means of providing guidance on the operation of the law. Section 16 of Health and Safety at Work etc. Act 1974, giving statutory force to this recommendation, provides that:

> For the purpose of giving practical guidance with respect to the requirements of any provision . . . the HSC may approve and issue such Codes of Practice (whether prepared by it or not) as in its opinion are suitable for that purpose.

The Health and Safety Executive typically explains the purpose of each code in the following way:

> [the] Code has been approved by the Health and Safety Commission and gives advice on how to comply with the law. This code has a special legal status. If you are prosecuted for breach of health and safety law, and it is proved that you have not followed the relevant provisions of the Code, a court will find you at fault, unless you can show that you have complied with the law in some other way.

Though an approved code may only be advisory it is nevertheless legally significant. A failure on the part of someone to observe any provision of an approved code of practice does not of itself render that person liable to civil or criminal proceedings (s.17). However if the advice contained therein is followed this will be prima facie evidence of compliance with the relevant provision. Similarly non-compliance with the code is prima facie evidence of non-compliance with the legislative requirement. In this case a potential offender will then have to show that he has undertaken works which are no less effective in securing compliance with the law. Section 17 of the Health and Safety at Work etc. Act 1974 provides that if the provisions of the code have not been observed a breach of the law is taken as proved unless the legal requirements have been 'complied with otherwise than by observance of the code'. It should be clearly understood that though the code does not have direct effect, in the sense that failure to observe the code is not an offence, a court, in any criminal proceedings, may, when considering evidence of any alleged contravention of a requirement or prohibition, have regard to the extent to which the code of practice was or was not adhered to. Evidence of a breach of a provision of a code of practice will have the effect, when taken with the provisions contained in s.40 of the Health and Safety at Work etc. Act 1974 (dealing with the burden of proof), of obliging an accused person to prove that what was done was all that was practicable or reasonably practicable. The codes supplement the statutory provisions. They transpose the often obscure language of the law into a more accessible form – a form which all interested parties can more readily understand. It is in the codes that obligations are clarified and the purpose of the law is more clearly revealed.

Though they are not always of the same legal significance as in the area of health and safety at work, nevertheless codes of practice are of the utmost relevance to the work of Environmental Health Officers. They are a source of information to which EHOs increasingly must have regard when interpreting and applying the law. Section 40 of the Food Safety Act 1990 makes provision for the issuing of codes of practice regarding the execution and enforcement of that Act. Food authorities, and therefore Environmental Health Officers, are obliged to have regard to the provisions of such codes and must comply with directions to take specified steps contained in those codes.

Department of the Environment Circular 17/96 is clear that for the assessment of unfitness under the Housing Act 1985 an Environmental Health Officer may have to have regard to a number of codes including the following:

BS 5262  Code of Practice for External Renderings
BS 5534  Code of Practice for Slating and Tiling
BS 6399  Loading for Buildings, Part 1:1996 Code of Practice for Dead and Imposed Loads
BS 8102  Code of Practice for Protection of Structures against Water from the Ground
CP 102:  1973 Protection of Buildings against Water from the Ground
BS 5720: 1979 Code of Practice for Mechanical Ventilation and Air Conditioning in Buildings
BS 5250: 1989 Code of Practice: the Control of Condensation in Buildings

BS 8303: 1986 Code of Practice for Installation of Domestic Heating and Cooking Appliances Burning Solid Mineral Fuels

BS 6465 Sanitary Installations Part 1: Code of Practice for Scale of Provision, Selection and Installation of Sanitary Appliances

BS 8206: Lighting for Buildings, Part 2 Code of Practice for Daylighting

BS 5572: 1978 Code of Practice for Sanitary Pipework

BS 6465 Sanitary Installations, Part 1 1994, Code of Practice for Scale of Provision, Selection and Installation of Sanitary Appliances

BS 6367: 1983 Code of Practice for the Drainage of Roofs and Paved Areas

BS 8301: 1985 Code of Practice for Building Drainage.

A failure to appropriately consider such guidance might legitimately lead to the Environmental Health Officer's opinion on unfitness being successfully challenged. It is certain that, in an entirely different area, that of noise control, a failure to comply with the code of practice on noise control on construction sites would be something which a court would take account of in any proceedings. Section 71 of the Control of Pollution Act 1974 provides for the issuing or adopting of codes of practice for matters under Part III of that Act. Section 60 of the Act enables a local authority to control noise emanating from construction operations. The authority only needs to be satisfied that the works to which the section applies (erecting, constructing, altering, repairing or maintaining buildings, structures or roads, or breaking up, opening or boring under any road or adjacent land in connection with construction, inspection, maintenance or removal works or any demolition or dredging work, or other work of engineering construction) are being or are going to be carried out on any premises. Where works falling within this section are being or are going to be executed on any premises the local authority may serve a notice imposing requirements as to the way in which the works are to be carried out. However when taking such a course an authority is bound to have regard to any code adopted under s.71. The Control of Noise (Code of Practice for Construction and Open Sites) Order 1984 designates BS 5228 Control of Noise on Construction and Open Sites Parts 1, 2, 3, 4 1984 as an approved code and thus it is obligatory on a local authority to have regard to this. A failure to do so will certainly render any decision by the authority subject to challenge.

# Basic criminal law 2

## 2.1 AN OUTLINE OF CRIMINAL LIABILITY

In this book we are very much concerned with the role of the Environmental Health Officer in the criminal law – his role as 'policeman, regulator or enforcer' involved in the detection and prosecution of crime. Yet even when adopting this limited approach there are problems. The first problem encountered when examining the role of the Environmental Health Officer in the criminal law is that there exists no substantive definition of what a crime is. There is no easily assimilated marker which can act as an unfailing check by which to determine whether given conduct is criminal. Perhaps the best approach is to adopt that offered by many writers in the field and say that a crime is conduct capable of being followed by criminal proceedings, having a criminal outcome. Further, proceedings or their outcome are criminal if they have certain characteristics which mark them as criminal.[1] Many of the statutory provisions enforced by Environmental Health Officers answer this description and thus come within the definition.

### 2.1.1 THE DISTINCTION BETWEEN CRIMINAL AND CIVIL PROCEEDINGS

The basic difference is that criminal proceedings are normally, though not exclusively, handled by an agency of the state: the Crown Prosecution Service, the Health and Safety Executive, a local authority etc. Such agencies will prosecute matters before the criminal courts, and are required to prove their case to the criminal standard i.e. beyond reasonable doubt. The wrongdoer here is termed the accused or the defendant and there are often 'victims' of the crime. Because a criminal prosecution is brought with a public interest element present then an individual (the victim) is not simply permitted to discontinue that prosecution at his discretion; it is not after all he or she who is prosecuting but the state. The aim of proceedings in criminal matters is that a court shall punish the offender by the imposition of a criminal sentence of imprisonment, fine etc. on him. The consent of the victim to a criminal offence is irrelevant, once again because of the public interest in the matter. I cannot consent to another killing me, even if I am a full supporter of euthanasia. Any person who therefore helps me in my death will be guilty of murder or manslaughter, despite my consent.

By contrast in civil proceedings the action is brought by one individual, who has been wronged, suing another in the civil courts. They will have to prove their case to the civil standard: the balance of probabilities. Such an individual can discontinue his action at any time and where the victim has consented to the wrong no action will lie as this will provide a valid defence. The aim in civil matters is to redress a grievance or to obtain compensation (damages) from the other and not primarily to obtain punishment. The parties are the plaintiff and the defendant.

It is important to remember that for any single act there may be both criminal and civil liability. And the nature of the act alone is insufficient to indicate which. For example an accident which results in injury to a worker may give rise to criminal proceedings under the Health and Safety at Work etc. Act 1974 and to a civil suit in the tort of negligence. A breach of s.14 of the Food Safety Act 1990, selling food not of the nature etc., may be both a criminal offence and a breach of contract. Conduct threatening physical eviction of a tenant of a house over a prolonged period may be both a breach of the covenant for quiet enjoyment implied into a tenancy agreement and a criminal offence under s.1(3) of the Protection from Eviction Act 1977.

To establish the existence of criminal conduct it is necessary to examine the nature of the circumstances which have occurred, having regard to the definition of the particular crime in question.

## 2.2 PROOF

If there is one essential point to remember in the criminal law it is that 'near enough is not good enough'. A nice, short phrase which in fact summarises neatly the position with regard to what an Environmental Health Officer, engaged in enforcement activities, may have to do. He will have to prove his case in the criminal courts beyond a reasonable doubt, not near enough beyond reasonable doubt but beyond that doubt. To prove it 'near enough' will not be to prove it 'good enough'. Further the Environmental Health Officer must prove all elements of the offence beyond that doubt and if he fails to prove all those elements, he fails to prove that offence. There is generally no burden on an accused person to prove anything whatsoever.

In a criminal case the prosecution must prove that all physical elements constituting the offence charged, the *actus reus*, were brought about by the conduct of the defendant. Where there is more than one element to the offence in question each must in turn be proved. In addition the prosecution must prove that the defendant created or brought about those elements in a particular state of mind, the *mens rea*. A failure to prove these matters beyond reasonable doubt will result in an acquittal.

### 2.2.1 THE CRIMINAL BURDEN OF PROOF

As already stated, where the prosecution is obliged to prove matters it must ordinarily do so beyond a reasonable doubt. The classic exposition of the burden on the prosecution is to be found in the case of *Woolmington v DPP* [1935]. In this rather sad case, Woolmington was charged with murdering his estranged wife. He had visited his wife and threatened to shoot himself if she did not return to him. In an ensuing struggle she was shot and killed by a bullet fired from the rifle he was holding. In denying the accusation of murder Woolmington claimed he squeezed the trigger involuntarily whilst threatening to shoot himself. Woolmington admitted manslaughter but denied murder. The judge at the trial, in summing up, told the jury that so long as the prosecution had shown that the accused had caused the death malice was presumed and that the accused must prove that the death was an accident. Woolmington was convicted of murder. On appeal to the House of Lords Lord Sankey delivered his judgment on the 'Golden Thread of English Law'. He stated:

> Throughout the web of the English criminal law one golden thread is always to be seen, that it is the duty of the prosecution to prove the prisoner's guilt .... If at the end of and on the whole of the case, there is a reasonable doubt, created by the evidence given by either the prosecution or the prisoner, as to whether the prisoner killed the deceased with a malicious intention, the prosecution has not made out the case and the prisoner is entitled to an acquittal. No matter what the charge or where the trial, the principle that the prosecution must prove the guilt of the prisoner is part of the common law of England and no attempt to whittle it down can be entertained.

There can therefore be no attempt to 'whittle down' the requirement that the prosecution should prove beyond reasonable doubt. But what is reasonable doubt? In this context it has been authoritatively stated that proof beyond reasonable doubt means 'such a degree of certainty that a man would seek to reach on an important decision in the affairs of his own life'.

Denning J in *Miller v Minister of Pensions* [1947], on the matter of proof beyond reasonable doubt stated:

> That degree is well settled. It need not reach certainty, but it must carry a high degree of probability. Proof beyond reasonable doubt does not mean proof beyond the shadow of a doubt. The law would fail to protect the community if it admitted fanciful possibilities to deflect the course of justice. If the evidence is so strong against a man as to leave only a remote possibility in his favour, which can be dismissed with the sentence 'of course it is possible but not in the least probable' the case is proved beyond reasonable doubt, but nothing short of that will suffice.

In *R v Summers* [1952], the approach to the standard for a jury was to be 'satisfied so that that they feel sure'. Lord Goddard CJ said:

> If a jury is told that it is their duty to regard the evidence and see that it satisfies them so that they can feel sure when they return a verdict of guilty, that is much better than using the expression 'reasonable doubt' and I hope in future that will be done.

Thus an Environmental Health Officer who hopes for a conviction because he has evidence on which a court can be 'pretty certain' (*R v Law* [1961]) or 'reasonably sure' (*R v Head* (1961)) of an accused's guilt, will fail in his task. If the Environmental Health Officer thinks he has done enough to establish that an accused is in all probability guilty, he will fail. For it is only when a criminal court gets beyond this, beyond 'in all probability', beyond reasonable doubt, that it can convict.

The approach of an Environmental Health Officer should thus be to gather and present evidence which will prove matters beyond reasonable doubt and enable magistrates or a jury to be satisfied that they are sure of guilt (*Ferguson v R* [1979]).

## 2.2.2 THE CIVIL STANDARD OF PROOF

This is often said to be the 'ordinary' standard and is usually stated as being that the court must be satisfied 'on the balance of probabilities'. Sometimes it is said that the court must be satisfied on a preponderance of probability, or that is 'it is more likely than not'. In American civil cases the expression a 'preponderance of the evidence' may be used. In *Miller v Minister of Pensions* [1947] Lord Denning said of the evidence required to satisfy the standard that:

> It must carry a reasonable degree of probability, but not so high as is required to discharge a burden in a criminal case. If the evidence is such that the tribunal can say: 'We think it more probable than not', the burden is discharged, but if the probabilities are equal it is not.

Clearly this is, and is meant to be, a lesser standard than that required in a criminal matter and imposes a consequently lesser burden on the person trying to establish a given fact.

### 2.2.3 BURDEN ON AN ACCUSED IN CRIMINAL MATTERS

Generally it is true to say that an accused person in a criminal matter has no burden of proof upon them, the phrase 'he who asserts must prove' generally being a good guide to the situation. However this is only a general rule and there are exceptions. If these exceptions apply then an accused may find himself subject to a burden of proof. These exceptions, which are effectively statutory, may be classified under two headings: when the statute imposes a burden on an accused by implication and when it expressly does so.

#### 2.2.3.1 The Burden by Implication

Section 101 of the Magistrates Court Act 1980 deals with the situation where the onus of proving a defence is implied on a defendant. This section provides that where:

> the defendant to an information or complaint relies for his defence on any exception, exemption, proviso, excuse or qualification, whether or not it accompanies the description of the offence or matter of complaint in the enactment creating the offence or on which the complaint is founded, the burden of proving the exception, exemption, proviso, excuse or qualification shall be on him, and this notwithstanding that the information or complaint contains an allegation negativing the exception, exemption, proviso, excuse or qualification.

Regrettably the precedents in this area lead one to the conclusion that it is unclear when a defendant will be under such a burden to prove. Following the decision of *R v Hunt* [1987] it would appear that any statutory provision in question must be construed individually to see if 'by necessary implication' the burden on an accused is present. *Hunt* concerned a charge of possessing a controlled drug contrary to s.5(2) of the Misuse of Drugs Act 1971. The House of Lords held that there was no rule of law that the burden of proving a statutory defence lay on the accused only where a statute so provided expressly. A statute might also place the burden of proving a defence on an accused by 'necessary implication'. And though each case will turn on the construction of the relevant statute the courts should be extremely slow to infer that a burden of proof was so imposed. What would appear to be relevant would be how onerous such a burden would be if placed on an accused. The more onerous the burden, the less likely will it be that it is required.

For example it is an offence under section 80(4) of the Environmental Protection Act 1990 to contravene or fail to comply, without reasonable excuse, with any requirement contained in a notice served under s.80. Such a person will therefore face an allegation that they did without reasonable excuse fail to comply with an Abatement Notice pursuant to section 80 of the Environmental Protection Act 1990 contrary to section 80(4) of the Act. The question arises: is the lack of a reasonable excuse part of the offence, and therefore for the prosecution to prove, or is non-compliance with the notice all that is required for the offence, with the defendant being obliged to establish reasonable excuse? Prima facie non-compliance with the notice is unlawful. Additionally the burden on an accused to show reasonable excuse is relatively light when compared with the burden on the prosecution, were it to have to show a lack of a reasonable excuse. In those circumstances the burden is on the accused. In *A Lambert Flat Management Ltd.* v *Lomas* [1981] the managers of a block of flats were prosecuted and convicted of failing to comply with a notice served under s.58 of the Control of Pollution Act 1974 (which contained provisions similar to those now found in s.80 of the Environmental Protection Act 1990). They had neither appealed nor complied with the notice.They alleged that an implied consent by the tenants of the flats to the nuisance was a valid defence, the notices were not therefore justified and they had a 'reasonable excuse' for non-compliance. The Crown Court allowed the appeal. In the Divisional Court Skinner J indicated what a reasonable excuse might be. He said:

> What ... if the person liable did not receive the notice because he was ill or abroad? The answer ... is that these are matters that can and would constitute a reasonable excuse ...

Clearly it would be easier for an accused to establish that they were ill or abroad than it would be for the prosecution to prove that they were not. In the case of *Wellingborough D.C. v Gordon* (1990) the court found that a birthday celebration was not a reasonable excuse for contravening a noise abatement notice. While the defence established that such circumstances as previous good behaviour, frequency, intensity and time of night of the noise could be taken into consideration in deciding whether there was a nuisance and while actions taken to minimise the expected impact and irritation caused by the noise might be matters of mitigation, the court found that they did not amount to a 'reasonable excuse' sufficient for a defence. Clearly here it was accepted that it was for the defence to establish reasonable excuse.

In *Nimmo v Alexander Cowan & Sons Ltd.* [1968] the House of Lords considered on whom the burden of proof fell when an offence contrary to s29(1) of the Factories Act 1961 was committed. The section provided that working places 'shall so far as is reasonably practicable be made and kept safe for any person working there'. The House of Lords found that it was for the defendants to prove that it was not reasonably practicable for them to provide and maintain a safe working environment. The court found that the prohibited act was that of providing a place of work which was not safe. Such an employer, to avoid liability, had then to prove that he had done all that was reasonably practicable.

### 2.2.3.2 The Burden Expressly Applied

The justification for finding the approach in the *Nimmo* case correct is surely to be found in the fact that s.40 of the Health and Safety at Work etc. Act 1974 now provides that if an offence under that Act is one which consists of a failure to comply with a duty or requirement to do something so far as is practicable or so far as is reasonably practicable, or to use the best practicable means to do something, it is for the defence to prove that it was not practicable or not reasonably practicable to do more than was in fact done to satisfy the duty or requirement, or that there was no better practicable means than was in fact used to satisfy the duty or requirement.

Under s.7(4) of the Environmental Protection Act 1990 there is a general condition to be implied in every authorisation under Part I. This condition requires that in carrying out the process to which an authorisation applies, the person carrying it out will use the Best Available Technique Not Entailing Excessive Cost (BATNEEC). If a person carrying on a prescribed process fails to use BATNEEC he will be in breach of the implied condition of the authorisation and thus commit an offence. By virtue of s.25(1) of the Environmental Protection Act 1990, in any prosecution involving an allegation that there has been a failure to comply with the general condition implied by s.7(4) it shall be for the accused to prove, on a balance of probabilities, that there was no better available technique not entailing excessive cost than was in fact used.

Section 40 of the Health and Safety at Work etc. Act 1974 and s.25 of the Environmental Protection Act 1990 are clear examples of when a statute expressly imposes a burden on an accused. They and other similar provisions do not operate so as to make an accused guilty until proven innocent. They still require the prosecution to establish beyond a reasonable doubt those matters which constitute the offence but once proven then the accused will find themselves required to prove they had done, in these cases, all that was reasonably practicable or applied BATNEEC.

Where there is a burden of proof placed on an accused in a criminal matter, whether expressly or by implication, that burden is to prove only to the civil standard i.e. on the balance of probabilities (*R v Carr-Briant* [1943]). Thus a person accused under s.2 of the Health and Safety at Work etc. Act 1974 need only prove, on the balance of probabilities, that they did that which was reasonably practicable. It is important to be clear that this does not require a person charged with an offence under the Act to prove their innocence, it only requires them to prove that they had done all that was reasonably practicable. Anyone so charged is not however required to introduce any such proof until the prosecution has established a prima facie case against them. Once that prima facie case has been established the burden shifts and it is for those accused to prove that it was not reasonably practicable for them to do more to fulfil their duty of care.

## 2.3 CRIMINAL LIABILITY

To establish criminal liability the prosecution is obliged to establish two things: that the elements constituting the offence charged were brought about by the conduct of the defendant (*actus reus*); that the defendant brought about those elements with a particular state of mind (*mens rea*).

For the first part of this it is vital to look to those elements which are to be found in the statutory or common law definition of the offence. The basic position of the criminal law is often summarised in the Latin maxim, *Actus non facit reum nisi mens sit rea*, an act does not make a man guilty of a crime unless his mind be also guilty. The use of Latin in this area is not without its critics. In *R v Miller* [1983] Lord Diplock in the House of Lords disapproved of such Latin axioms, stating that:

> … it would I think be conducive to clarity of analysis of the ingredients of a crime that is created by statute, as are the great majority of criminal offences today, if we were to avoid bad Latin and instead to think and speak … about the conduct of the accused and his state of mind at the time of that conduct, instead of speaking of *actus reus* and *mens rea*.

Nevertheless the use of these expressions in this area of the law is so deeply embedded and widespread that the Environmental Health Officer would be wise to simply accept and use them until the future brings change.

### 2.3.1 ACTUS REUS: THE EXTERNAL OR OBSERVABLE ELEMENTS

*Actus* literally translates to mean 'act', and though most crimes do require that the defendant commit a certain act this is not always the case. In *Haughton v Smith* [1975] Lord Hailsham said:

> I desire to make an observation on the expression *Actus Reus* ... Strictly speaking, though in almost universal use, it derives, I believe, from a mistranslation of the Latin aphorism: *Actus non facit reum nisi mens sit rea.* Properly translated this means 'An act does not make a man guilty of a crime unless his mind be also guilty.'

Lord Hailsham here was alluding to the fact that criminal liability may arise not simply out of an act but also through a failure to act. Thus to translate *actus* simply as relating to an act is to fail to recognise that a failure to act, an omission, may well constitute the *actus reus* of an offence. Most commentators have come to the conclusion that the word 'conduct' is preferable. Conduct can include both acts and omissions. So qualified, the words *actus reus* can be considered as the physical elements, external elements, or, as it has sometimes been referred to, the 'observable doing' of a crime. This is in contrast to the mental or internal elements of an offence (*mens rea*).

A crime will normally be defined by reference to more than mere conduct. Typically a crime will be composed of

(a) some conduct
(b) certain surrounding circumstances and (sometimes)
(c) a prohibited consequence.

These elements are to be found in the description of the offence contained, usually, in the statute. Take for example s.1 of the Criminal Damage Act 1971. If the section is examined it is found that the statute provides that a person who without lawful excuse destroys or damages any property belonging to another intending to destroy or damage any such property or being reckless as to whether any such property would be destroyed or damaged, shall be guilty of an offence.

The external elements, *actus reus*, of this offence are, without lawful excuse, destroying or damaging property belonging to another. The conduct is destroying or damaging property, the surrounding circumstances are that the property belongs to another and there is no lawful excuse.

Property may in certain cases thus be destroyed without committing any offence. For example, if a person burns down his house in order to claim insurance, there may be an offence of dishonesty, but there can not be an offence under section 1(1). The prosecution could not establish that he has damaged or destroyed property belonging to another. Section 1 of the Theft Act 1968 provides that a person is guilty of theft if he dishonestly appropriates property belonging to another with the intention of permanently depriving the other of it. Thus to appropriate one's own property is to commit no offence, as one of the circumstances of the *actus reus* is missing. The *actus reus* of theft is the appropriation of property belonging to another.

Section 80(4) of the Environmental Protection Act 1990, as has been seen, makes it an offence for a person, on whom an abatement notice has been served, to, without reasonable excuse, contravene or fail to comply with the requirements or prohibitions imposed by that notice. The conduct here is failing to comply with the requirements or prohibitions imposed by the notice, the surrounding circumstances are that there is no reasonable excuse. Using this method of analysis it has been said that what constitutes the *actus reus* of an offence are acts (including omissions) committed in legally relevant circumstances causing a prohibited result[2].

### 2.3.2 PROOF OF ACTUS REUS

Whatever the crime the prosecution must prove that every part of the *actus reus* was in fact done by the accused. A failure to do so will result in an acquittal even if the conduct may appear morally wrong. In the case of *R v Deller* (1952) the defendant sold a car which he believed to be subject to a hire-purchase agreement with payments outstanding. The nature of a higher purchase agreement is such that ownership in the goods does not pass to the purchaser until the last payment is made, therefore to sell the goods subject to the agreement is to sell property belonging to another. In fact the car was free from all encumbrances. Deller was subsequently charged with and convicted of obtaining property by deception. It is established that there can be no deception unless a person is induced to believe that a thing is true which is in fact false (*Re London and Globe Finance Corp. Ltd.* [1903]). Thus, on appeal, the conviction was quashed as the car was free from any encumbrance. Deller, perhaps unintentionally, had not deceived the purchaser. The purchaser believed the car to be free of encumbrances, he was not induced therefore into believing a thing to be true which was in fact false, so the *actus reus* of deception was absent. Similarly in *R v White* (1910) the defendant had put 2 grains of potassium cyanide in his mother's drink. The mother subsequently died, but of a heart attack not of poisoning. A charge of murder could not be sustained unless it was the poison which killed her. As the deceased died of a heart attack there was only the *actus reus* of attempted murder but not of murder.

### 2.3.3 OMISSIONS

As a general principle there is no criminal liability for a failure to act. However, the law has set out certain factual situations in which persons shall be under a legal duty to act, and if they fail to act in these situations thereby causing a prohibited criminal result, they shall be liable for that result. Put another way the *actus reus* of an offence can be a failure to act.

Under the Health and Safety at Work etc. Act 1974 one of the principal offences (section 33) is a failure to carry out one or more of the general duties imposed under sections 2 to 7 of the Health and Safety at Work etc. Act 1974. Under s.23(1)(c) of the Environmental Protection Act 1990 there is the offence of failing to comply with enforcement or prohibition notices. Similarly under the Food Safety Act 1990, s.10(2) creates the offence of failing to comply with an improvement notice. In each of these cases the offence will involve not the act of an accused but his failure to act. It is also possible for individuals to be liable for a failure to act where they are obliged so to do by contract (*R v Pittwood* (1902))   or where there has been a voluntary assumption of responsibility (*R v Nicholls* (1874)).

### 2.3.4 THE INTERNAL ELEMENTS: MENS REA

The expression *mens rea* is generally translated as the 'guilty mind', it may also be known as the guilty intent. Such a translation, though useful shorthand, is, as we have already seen with regard to *actus reus*, too imprecise and misleading an expression. It is better and more conducive to understanding to merely consider it to be the mental (or internal) element of the crime. This approach is better as it is possible for an accused person to have committed an offence without any obviously 'guilty' mind. *Mens rea* is that term which is given to a defendant's state of mind and which is required to be proved in relation to each of the elements of the *actus reus* of an offence. Firstly it is worth remembering that ignorance of the law is no excuse; as knowledge of the law is not a constituent of *mens rea* it is accordingly no defence that an accused was unaware of the law. However, to establish quite what the state of mind that is required to be proved, is we must look to the definition of the offence. The state of mind required for an offence will vary according to the definition of the offence . However, generally, there are a number of states of mind recognised. These may be represented in terms of increasing culpability by Figure 2.1 and will be looked at in turn.

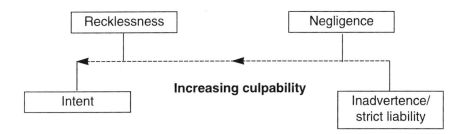

**Figure 2.1** States of mind

## 2.3.5 INTENT

This is the most blameworthy or guilty of the states of mind and is often encountered by the Environmental Health Officer. For example s.33 of the Food Safety Act 1990 creates the offence of obstruction and provides that any person who intentionally obstructs any person acting in the execution of the Act shall be guilty of an offence. The state of mind which must accompany the *actus reus* of obstruction is intent. Unfortunately there is no agreement in decided cases as to the precise meaning of intent, nor is it defined by any statute.

In *Hyam v DPP* [1975] the House of Lords considered the meaning of intent with regard to murder. In this case Pearl Hyam was jealous of her lover's new mistress, one Mrs Booth. She wished to frighten Booth into leaving the area. In the early hours of the morning she poured petrol through Booth's letter box and set it alight. Mrs Booth and one of her children escaped but two of the children died in the ensuing fire. The question before the House of Lords was could Hyam be convicted of murder if she did not directly intend to cause death or grievous bodily harm, but foresaw death or grievous bodily harm as a highly probable result of her actions. Lord Hailsham in his judgment referred to the case of *Cunliffe v Goodman* [1950] and said:

> I know of no better judicial interpretation of 'intention' or 'intent' than that given in a civil case by Asquith L.J. when he said:
>
> > 'An intention', to my mind, connotes a state of affairs which the party 'intending' ... does more than merely contemplate. It connotes a state of affairs which, on the contrary, he decides, so far as in him lies, to bring about and which, in

point of possibility, he has a reasonable prospect of being able to bring about, by his own act of volition.

... I think it is clear that 'intention' is clearly to be distinguished alike from desire and foresight of the probable consequences.

Lord Hailsham stated that he did not believe that knowledge or any degree of foresight is enough. Knowledge or foresight is at best material which entitles or compels a jury to draw the necessary inference as to intention. Foresight does not necessarily imply the existence of intention. Foresight of consequences is no more than evidence of the existence of intention. In *R v Moloney* (1985) the accused and his stepfather had been drinking heavily during a ruby wedding celebration. There followed a disagreement as to who was fastest at loading and firing a shotgun. This resulted in a race which the accused won. The stepfather was reported to have taunted the accused by saying he had no nerve to pull trigger of a loaded barrel. The accused pulled that trigger and killed the stepfather. The accused later stated that he had not aimed the gun, 'I just pulled the trigger and he was dead'. The House of Lords held that the *mens rea* for murder was intention to kill and foresight of probability of death was not the same as intention. In *R v Hancock and Shankland* [1986] the House of Lords found that foresight of consequence is not to be equated with intention – foresight of probability or even certainty was not intention. It is impossible for anyone to truly know what goes on inside the head of another and therefore the state of mind of an accused must be inferred from the circumstances surrounding an offence.

Section 8 of the Criminal Justice Act 1967 provides that a court or jury

(a) shall not be bound . . . to infer that [an accused] intended or foresaw a result of his actions by reason only of its being a natural and probable consequence of those actions; but
(b) shall decide whether he did intend or foresee that result by reference to all the evidence drawing such inferences from the evidence as appear proper in the circumstances.

Although the cases so far considered relate to murder, in *AMK (Property Management) Ltd.* [1985] the principles found in these cases were applied with equal force to an offence more commonly encountered by Environmental Health Officers, that of an offence under section 1(3) of the Protection from Eviction Act 1977. This Act protects residents and occupiers against both unlawful eviction and harassment. Under s.1(3) it is an offence if any person with intent to cause the residential occupier of any premises either:

(a) to give up the occupation of the premises or any part thereof; or

(b)to refrain from exercising any right or pursuing any remedy in respect
  of the premises or part thereof

does acts likely to interfere with the peace or comfort of the residential
occupier or members of his household, or persistently withdraws or
withholds services reasonably required for the occupation of the
premises as a residence. In fact because of difficulties associated with the
proof of intent, s.29 of the Housing Act 1988 introduced an offence of
harassment without the need to prove intent.

It is an offence under section 1(3A), which was added by the 1988 Act,
for the landlord of a residential occupier or his agent to:

(a)do acts likely to interfere with the peace or comfort of the residential
  occupier or members of his household; or
(b)persistently withdraw or withhold services reasonably required for
  the occupation of the premises as a residence

and in either case he knows or has reasonable cause to believe that the
conduct is likely to cause the residential occupier to give up the occupation of the whole or part of the premises or to refrain from exercising any
right of pursuing any remedy in respect of the whole or part of the
premises.

## 2.3.6 RECKLESSNESS

Under s.33 of the Health and Safety at Work etc. Act 1974 it is an offence
to intentionally or recklessly interfere with anything provided for safety
(s.8). It is also an offence for a person to make a statement which he
knows to be false or recklessly to make a statement which is false where
the statement is made:

i) in purported compliance with a requirement to furnish any information imposed by or under any of the relevant statutory provisions; or
ii) for the purpose of obtaining the issue of a document under any of the
    relevant statutory provisions to himself or any other person.

As for 'intent' the legal interpretation of recklessness suffers from a lack
of clarity. In the criminal law recklessness has two meanings:

(a) a deliberate taking of an unjustified risk
(b) giving no thought to the existence of an unjustified risk.

This rather unusual situation has come about as the result of two separate cases: *R v Cunningham* [1957] and *R v Caldwell* [1982] .

In *Cunningham* the defendant was accused of an offence contrary to
s.23 of the Offences Against the Person Act 1861, which involved a
charge of unlawfully and maliciously administering or causing to be

administered to any other person any poison or other destructive or noxious thing so as to endanger the life of such a person. The accused here had removed a gas meter and taken money from it. This caused gas to leak to an adjoining property and partially asphyxiate someone. The judgment by the Court of Appeal in *Cunningham* approved an earlier statement of the law by Professor Kenny in the first edition of his *Outlines of Criminal Law* published in 1902:

> In any statutory definition of a crime, malice must be taken ... as requiring either (1) an actual intention to do the particular kind of harm that in fact was done; or (2) recklessness as to whether such harm should occur or not (i.e. the accused had foreseen that the particular kind of harm might be done and yet had gone on to take the risk of it).

Recklessness here means the conscious taking of an unjustifiable risk. It is a subjective test. The second type of recklessness is known as *'Caldwell'* type recklessness and carries elements of a more objective approach. The accused was charged with arson contrary to s.1(1) and (2) of the Criminal Damage Act 1971. At his trial he pleaded guilty to the lesser charge of intentionally or recklessly destroying or damaging any property belonging to another contrary to s.1(1), but not guilty to the more serious charge under s.1(2) of damaging property with intent to endanger life.

Lord Diplock in the House of Lords said:

> In my opinion, a person charged with an offence under section 1(1) of the Criminal Damage Act 1971 is 'reckless as to whether any such property would be destroyed or damaged' if (1) he does an act which in fact creates an obvious risk that property will be destroyed or damaged and (2) when he does the act he either has not given any thought to the possibility of there being any such risk or has recognised that there was some risk involved and has nonetheless gone on to do it. That would be a proper direction to the jury.

Applying this not simply to recklessness in connection to damage to property, it is reckless where an accused person does an act which creates an obvious risk that the prohibited consequence will occur and he either has not given any thought to the possibility of there being any such risk or he has recognised that there was some such risk and yet has carried on with his actions. Under this test the risk need only have been obvious to a reasonably prudent adult. It is still uncertain to which crimes the different definitions apply. However the current position appears to be that if the statutory definition of a crime contains the expression recklessness then that is to be construed as *Caldwell* type reck-lessness. But not if it does not.

### 2.3.7 NEGLIGENCE

In *Blythe v Birmingham Waterworks* (1856) negligence was said to be the omission to do something which a reasonable man guided upon those considerations which ordinarily regulate the conduct of human affairs would do, or doing something which a prudent and reasonable man would not do. The Law Commission in Working Paper no. 31 said that: 'a person is negligent if he fails to exercise such care, skill or foresight as a reasonable man in his situation would exercise'.

Negligence as a state of mind forms the basis of very few crimes, arguably the best known of which is careless driving contrary to s.3 of the Road Traffic Act 1988, though it may be found in some environmental statutes. For example s.73 of the Water Industry Act 1991 provides that an owner or occupier of premises to which water is supplied commits an offence if they intentionally or negligently cause or suffer a water fitting for which they are responsible to be so out of order that water in a main pipe of the water undertaker or any pipe connected thereto is contaminated or wasted. It remains true however that as a criminal state of mind negligence is not commonly encountered by the Environmental Health Officer.

## 2.4 OFFENCES OF STRICT LIABILITY

For some offences a person may be convicted in the absence of intention, recklessness or negligence. In such circumstances an offender is said to be under strict liability. Liability is here said to be strict because the need for *mens rea*, so far as at least one aspect of the *actus reus* is concerned, has been dispensed with. Offences of strict liability are almost exclusively the creation of statute. Thus it is essential to examine the relevant statute to ascertain whether or not it is the intention of Parliament to create an offence of strict liability. A failure to interpret the provision correctly can have very obvious consequences in terms of evidence accumulation and presentation.

One cardinal principle of statutory interpretation is that, in statutes creating criminal offences, there is a presumption that *mens rea* is required. This is so even where it is not specifically mentioned. In *Gammon (Hong Kong) Ltd v Attorney General of Hong Kong* [1985] part of a building in the process of construction collapsed and the defendants, who were the registered contractor, project manager and site agent, were charged with a breach of building regulations. The Privy Council on hearing the case held that the relevant regulations did create offences of strict liability. Lord Scarman summarised the position in this area of the law as follows:

1. There is a presumption of law that *mens rea* is required before a person can be held guilty of a criminal offence.
2. The presumption is particularly strong where the offence is 'truly criminal' in character.
3. The presumption applies to statutory offences and can be displaced only if this is clearly or by necessary implication the effect of the statute.
4. The only situation in which the presumption can be displaced is where the statute is concerned with an issue of social concern: public safety is such an issue.
5. Even where the statute is concerned with such an issue, the presumption of *mens rea* stands unless it can also be shown that the creation of strict liability will be effective to promote the objects of the statute by encouraging greater vigilance to prevent the commission of the prohibited act.

This presumption in favour of *mens rea* is rebuttable. That is to say it is a starting point for interpretation. As an assumption it is capable of being displaced. There are a number of factors which are likely to cause the presumption to be rebutted. Lord Scarman in *Gammon* speaks of offences 'truly criminal' in character, which is a reference to earlier decisions of the courts.

In *Sweet v Parsley* [1970], the House of Lords identified two types of criminal offence:

(a) those which could truly be said to be criminal;
(b) those which are not criminal in any real sense, but are acts which in the public interest are prohibited under a penalty.

In the first category the presumption in favour of *mens rea* will rarely be rebutted, whereas in the second category it is easier to infer that Parliament intended the presumption to be rebutted. A further indicator of the way the court might be expected to interpret a provision is the size of the penalty available. Generally the larger or more severe the penalty, the less likely the court is to treat it as a crime involving strict liability. In *Sweet v Parsley* [1970] the defendant was convicted of an offence under s.5(b) of the Dangerous Drugs Act 1965. The defendant had not known drugs were being used on her premises yet was nevertheless convicted. The House of Lords quashed this conviction and held that as the offence was 'truly criminal', with grave consequences for the defendant, it had to be proved that she had intended the house to be used for drug taking.

A court's first obligation is to give force to Parliament's intention as stated in the Act. It is not for the court to usurp the function of Parliament. And in certain cases having regard to the above and the wording of the statute the intention of Parliament may be clear. For example in regard to offences under s.14 of the Food Safety Act 1990 there is no need to prove *mens rea*. In *Betts v Armstead* (1888) it was held that an offence within what is now section 14 was committed even though the seller did not know that the article sold was not of the nature, substance or quality demanded. Cave J said:

> It was suggested that the justices ought to read into the section the word 'knowingly' ... That word is not to be found in this section, and it is clear from the words of other sections of the Act that the word 'knowingly was intentionally omitted.

In *Cundy v Le Coq* (1884) the court examined not just the section in question but other parts of the same Act to ascertain whether or not *mens rea* was required for the offence before them. The rationale behind this was that if words requiring *mens rea* are to be found in other sections of the Act then it may be presumed that where no such words are used this is a deliberate omission by Parliament meant to indicate the creation of an offence of strict liability. Cundy, a publican, was convicted of unlawfully selling liquor to a drunken person contrary to s.13 of the Licensing Act 1872. This particular statute provided that 'If any licensed person ... sells any intoxicating liquor to a drunken person he shall be liable to a penalty.' Though Cundy had sold the liquor he did not know the person concerned was drunk. Nevertheless it was held that knowledge of the condition of the person to whom the liquor was sold was not necessary to constitute the offence. The court examined other parts of the Act and found that the words 'knowingly' had been used elsewhere. Stephen J said:

> ... the substance of all the reported cases is that it is necessary to look at the object of each Act that is under consideration to see whether and how far knowledge is of the essence of the offence created. Here, as I have already pointed out, the object of this part of the Act is to prevent the sale of intoxicating liquor to drunken persons, and it is perfectly natural to carry out that by throwing on the publican the responsibility of determining whether the person supplied comes within that category.

In *Pharmaceutical Society of Great Britain v Storkwain Ltd.* [1986] the court was concerned with an offence of supplying prescription-only drugs contrary to s.58(2)(a) of the Medicines Act 1968. The defendant had provided prescription-only drugs to customers who had presented forged prescriptions, though he had believed, on reasonable grounds, that the prescriptions were genuine. The House of Lords found that there were, in other sections of the Act, express requirements for *mens rea*. The absence of such requirements in s.58(2)(a) thus led them to rebut the presumption that *mens rea* was required for the instant offence.

This approach is not however conclusive. In *Sherras v De Rutzen* (1895) a licensee supplied liquor to a police officer whilst on duty contrary to s.16(2) of the Licensing Act 1872 and was convicted by magistrates. On appeal, the court recorded that s.16(1) of the Act made it an offence for a licensee to 'knowingly' harbour or suffer to remain on his premises any constable on duty, while s.16(2) made no such reference to a state of mind. However this was held not enough to indicate that the offence contrary to s.16(2) was one of strict liability.

### 2.5 DEFENCES TO STRICT LIABILITY OFFENCES

A majority of the statutory offences involving strict liability arise under regulatory legislation aimed at public protection. The problem with offences which are of strict liability is that they may result in the punishment of both those who intentionally, recklessly or negligently perform the prohibited activity and those who, despite taking the greatest of care, do the same. This can appear harsh and even provide a deterrent to those who would wish to try their best. If the law does not differentiate between the careful and the careless then all may decide to become careless and thereby save the money that being careful might cost. Any public protection provision creating such a situation is obviously self defeating. There has to be some way of encouraging potential offenders to undertake work and expend money and effort to avoid liability. In fact the potential injustice and disincentive involved is reduced by the fact that for many offences of strict liability there are provided statutory defences. This perhaps helps to differentiate strict liability from absolute liability. These statutory defences operate in such a way as to allow the law to differentiate between the negligent harmdoer and the innocent offender.

Such defences tend to fall into two broad groups: 'no negligence' defences and 'third party' defences. A 'no negligence' defence involves the accused person establishing he was not negligent with regard to the commission of the offence. Such a defence is to be found in the Food Safety Act 1990. Section 21(1) of the Food Safety Act 1990 provides a general defence to anyone charged with any offence under Part II of the Act and requires that the defendant show that he took all reasonable precaution and exercised all due diligence. If a person is able to show that he took all reasonable precaution and exercised all due diligence to prevent the commission of an offence, he will be entitled to an acquittal. It is for the accused to establish, on the balance of probabilities, that all was done that should have been done. Obligations to exercise due diligence have been said to be indistinguishable from the obligation to exercise reasonable care. Therefore, to establish a defence under this section a defendant may only have to show that he acted without negligence, hence the expression a 'no negligence' defence. A conscientious food business proprietor who has tried his best to avoid committing an offence, i.e. he has been 'careful', may thus be entitled to use the defence. A similar though 'careless' proprietor may not. Here then is the stimulus to be careful and the reward for that care; that is the way that the provision avoids creating the disincentive indicated earlier.

The 'third party' defences require an accused not only to prove lack of negligence on his part but also that the contravention was due to the act (or default) of a third party. Such a third party might be a supplier or an employee. Again the Food Safety Act 1990 in s.21 shows an example of this type of defence. For offences under s.8, s.14, or s.16 of the Act the statutory defence is held to have been established if the accused satisfies the requirements found in s.21(3) or (4).

Section. 21(3) provides that it shall be a defence to show:

(a) that the commission of the offence was due to the act or default of another person, who was not under his control, or to reliance on information supplied by such a person;
(b) that he carried out all such checks of the food in question as were reasonable in all the circumstances, or that it was reasonable for him to rely on checks carried out by the person who supplied the food to him; and
(c) that he did not know and had no reason to suspect at the time of the offence that his act etc. would amount to an offence.

If both these things are proven by him the accused has a valid defence and any third party is then, in turn, liable to be prosecuted.

It is important to recognise that for any statutory defence all elements of that defence must be established for it to be effective. For example with a 'due diligence' defence it is useful to recall that in *Tesco Supermarket Ltd v Nattrass* [1972] Pearson L, in quoting the Divisional Court, said:

> The taking of such precautions and the exercise of such diligence involves, or may involve, two things. First the setting up of an efficient system for the avoidance of offences under the Act. Secondly it involves the proper operation of that system.

Lord Diplock added that what were reasonable precautions and the exercise of all due diligence was a matter of fact for the justices or the jury to decide. The defence here requires clearly two things. If an accused is able to establish only part of a statutory defence it will not afford any protection and the offender is likely to be convicted. In *Meah v Roberts* [1978] a fitter was employed to install and maintain equipment for the supply of draught beers. The fitter visited M's restaurant and cleaned the equipment using a caustic soda solution. The fitter did not use all of the solution and put what remained into a lemonade bottle. He altered the label by writing 'Cleaner' on. The bottle was then placed beneath the bar. There were other bottles of lemonade kept here. The fitter did not tell any member of the staff what he had done. M had not previously instructed his staff that cleaning materials and food were to be kept separate. A man visited the restaurant with his children. He ordered lemonade for the children who were, by mistake, given the caustic soda. M claimed a defence under section 100 of the Food Act 1984 (this dealt with the default of another coupled with exercise of all due diligence) on the basis that the offence had occurred through the act or default of the fitter and that he, M, had used all due diligence to prevent the offence. The court found both M and the fitter guilty of supplying food not of the nature demanded by the purchaser. M was unable to establish that he had used all due diligence when instructing his staff as to the storage of food and cleaning materials, i.e. he could not prove the second part of the defence on which he was seeking to rely.

A useful question to ask when confronted with the possibility of defences of this kind is: 'what more could the accused reasonably have done? If the answer is nothing, then it can be expected that the defence will be established. It is vitally important to remember to use the word 'reasonable'. In *Garret v Boots Chemist Ltd.* (1980) Lord Lane suggested that what might be reasonable for a large retailer might not be reasonable for the village shop. In *Texas Homecare Limited* v *Stockport Borough Council* (1987) the Divisional Court found:

> Reasonable diligence required the appellants to establish some kind of system . . . That does not mean that the system had to be foolproof, no system could be. Nor did the appellants have to examine every article. But they did have to do something.

It is important when considering such defence to avoid the problem of 20/20 hindsight. The computation of what is reasonable falls to be made by the potential offender at a point in time anterior to the incident and must be judged accordingly.

### NOTES

1 Williams G, 1955, The definition of Crime, *Current Legal Problems*, 107, 130.
2 Smith and Hogan, *Criminal Law: Cases and Materials* (2nd Ed 1980, p.31).

# Powers of entry 3

No man can set his foot upon my ground without my licence, but he is liable to an action, though the damage be nothing ... If he admits the fact, he is bound to show by way of justification that some positive law has empowered or excused him.

Lord Camden CJ Entick v Carrington (1765)

## 3.1 ENTRY ONTO LAND

There is an old common law maxim which states: *'cujus est solum ejus est usque ad coelum et ad inferos'* – To whom belongs the soil, his it is, even to heaven and to the middle of the earth. Though somewhat overstating the true legal position it nevertheless represents a reasonable indication of the extent of the benefits attaching to the ownership of land. Indeed much of English law is directly or indirectly concerned with the rights attaching to or derived from the ownership of land. One of these rights, and a key one jealously guarded, is the right to exclude a stranger from one's land. It follows from this that any unauthorised entry by a stranger can constitute a trespass. There is no requirement here to prove damage to property or goods. The wrong, the tort, is the unauthorised entry and trespass is thus said to be actionable 'per se' (of itself).

One of the most valuable and extensive powers available to an Environmental Health Officer is the power to enter premises. Much of the work of the Environmental Health Officer is indeed dependent on his being able to gain entry to premises to undertake routine inspections, investigations and surveys. The idea of risk assessment, increasingly at the heart of the work of the Environmental Health Officer, would be a barren concept if EHOs could not obtain access to premises. Because of the wide range of duties undertaken by the Environmental Health Officer, his range of powers of entry, search and seizure is equally extensive; perhaps more extensive than for any other public official in the country. Yet entering a man's house or place of business, searching it and seizing his goods against his will are tortious acts against which he is entitled to the protection of the court, unless that is, such acts can be justified either at common law or under some statutory authority (Lord Diplock in *R v IRC ex p. Rossminster Ltd* [1980]).

Routinely the Environmental Health Officer would expect to gain entry with the permission of the lawful occupier. This 'express permission' operates so as to ensure that the officer concerned does not enter the property as a trespasser and thus enters and remains on the property as a lawful visitor. However where anyone, and this includes an Environmental Health Officer, enters onto private property without permission, they do so, subject to what is discussed below, as a trespasser; they commit the tort of trespass to land and invite the possibility of an action for damages.

Any unauthorised entry by any person onto private premises is an actionable trespass, as is also remaining on land after permission to enter has been revoked or has expired. Thus where the Environmental Health Officer enters with permission and then is subsequently told to go, he or she must do so or leave him or herself, and perhaps their employer, open to an action for trespass.

Because a trespasser is considered to be an unwanted intruder, he may be removed from the premises by the use of reasonable force and his proposed entry may be resisted. In the case of *Stroud v Bradbury* [1952] Lord Goddard recorded that the landowner resisted the sanitary inspector with all the rights of a freeborn Englishman whose premises were being invaded and defied him with a clothes prop and spade. While the idea of a freeborn Englishman repelling an invader with a clothes prop and a spade may appear entertaining, in fact to allow such a situation to develop is unwise, unprofessional and simply dangerous. Situations such as this may all too easily escalate and result in both the Environmental Health Officer and the owner of the land becoming involved in a potentially violent confrontation. This can never be seen as a satisfactory outcome. It is therefore vital that Environmental Health Officers understand the limits of their right to be on land and equally the rights of occupiers to resist them.

Entry onto land may be lawful without there having been any express permission given. There may be an easement, a right of way, across the land of another which makes it perfectly lawful for a stranger to be on the land. Such an easement is a right attached to a piece of land (the dominant tenement) allowing the owner of that land (called the dominant owner) to use the land (the servient tenement) of another (the servient owner). This does not allow him to claim any ownership of the land nor to take any part of its produce or its soil. An easement is a right over land and not to it (*Copeland v Greenhalf* [1952]). Though the right to be on the land is here very restricted, nevertheless it will offer an adequate defence to an allegation of trespass. There may be an 'entry by licence', in which the permission may be implied on the part of the occupier. A licence is essentially a consent which, like an easement, without passing any interest in the property, prevents the act of entering onto property from being wrongful, i.e. a trespass. An example of an implied licence to enter property is that which a postman, paperboy or other necessary visitor may have. Against these no action for trespass would lie. There is taken to be an implied licence for them to enter onto the land, though for a very limited purpose. Obviously they acquire no interest in the land nor is a right of way created, yet their entry onto the land is lawful and no action in trespass lies. Similarly an Environmental Health Officer having legitimate business with the occupiers of a dwelling does not commit a trespass merely by opening a garden gate and walking to the front door. It is of course possible to expressly revoke such an implied consent by notices or verbal warnings. If this is done all affected visitors may legitimately be considered trespassers. The Environmental Health Officer may thus enter with permission, express or implied, but if at any point this permission is revoked, then he must remove himself from the premises or become a trespasser.

## 3.2 COMMON LAW AND STATUTORY ENTRY

Were the position to be simply as stated it would create some very real problems. The situation is not, in reality, so inflexible. The law has necessarily evolved to recognise two conflicting imperatives. The first is summed up by the overused and somewhat incorrect phrase 'an Englishman's home is his castle'. The second is the need to recognise that in any developed society having a mature legal system, there will be times when it is in the interests of that society to allow entry onto land without permission and without exposing the person so entering to the threat of action. So it is the case that the law recognises a number of situations when entry without permission is nevertheless not actionable, for example where it is necessary to save life or limb, or to prevent serious damage to property.

Such situations are however unlikely to be confronted by Environmental Health Officers. For them, where there is no express or implied permission and where the common law provisions do not apply, then the only other grounds for lawful entry are statutory. Here a specific statutory provision will provide for officers to enter private premises, which may be domestic or business, for the execution of specified activities. In granting this power the Acts do not seek to provide an umlimited right of access for all persons to all premises at all times. Any empowering Act will stipulate the conditions precedent for the exercise of a power of entry, by whom the power is exercisable and the limits of its application. Environmental Health Officers must therefore know all of the above if each particular provision is to be used.

3.2.1 AUTHORITY OF LAW AND AUTHORISATION

The idea, sometimes expressed, that all an Environmental Health Officer has to do, to obtain lawful entry to any premises, is to demand entry (possibly after having sought out a police officer to accompany him) is wrong. The power to enter onto private property is a significant, though necessary, restriction on the liberty and rights of others. It must therefore be expected that the law would require it to be undertaken carefully, precisely, and having full regard to the requirements of the authorising statute.

### 3.2.1.1 Authorisation

The Local Government Act 1972 provided for the first time a general power for local authorities to discharge any of their functions by delegation to officers. Thus every officer seeking to use any power of entry under any statutory provision must first be authorised by the relevant authority for that purpose. Guidance is produced by LACOTS (FS 7 95 2, May 1995) on the form such an authorisation might take under s.32 of the Food Safety Act 1990 and a number of the points raised there are of more general relevance.

An authorisation, which is not to be confused with the simple identification card which all officers ought to carry, should identify the local authority and the officer's department. It should identify the officer, by name and title, as an authorised officer and the provisions under which he is authorised. If the authorisation is not for all parts of a particular statute the details of the relevant parts or sections should be provided. This document should bear the signature of the officer and may often carry a photograph. It must be signed and dated by an appropriate person empowered to authorise the officer to act under the relevant statutory provisions; it may, for a local authority, sometimes carry the seal of the council. The card may have a registration number on it to enable verification of details on the card with departmental records and, to this end, the card may contain details of contact telephone numbers to allow verification by a member of the public. Finally the card may contain a reference as to where the card may be taken if found by a member of the public.

There is no national format for these documents and therefore their form and layout will vary across the country. However a typical format is reproduced for a general authorisation document (Form 3.1).

It must be emphasised how important it is that officers are clear on the extent of their authority and at all time carry such documentation with them. It must never be assumed that an officer has been authorised. Being employed to do a job is not the same as being authorised for the

## General Authorisation

**It is hereby authenticated that.....................................
the bearer of this document, is a duly authorised officer of the
Anytown City Council under the statutory provisions contained in
the schedule attached hereto. The officer is further authorised
under those provisions to enter premises and to examine items
thereon as provided for under, and in accordance with, those
enactments in the exercise and discharge of the functions of the
Council.
I hereby also authenticate the signature and
photograph of the said......................................**

| Signature |
| :---: |
| ......................... |
| **PHOTOGRAPH** |

**Dated this ...............day of....................... 19......**

**Signed for and on behalf of Anytown City council**

.................................................................

**REFERENCE NUMBER (To be used for all telephone enquiries)...........**

### Schedule of Statutory Provisions

Animal Boarding Establishments Act, 1963
Animal Health and Welfare Act, 1984
Breeding of Dogs Act, 1973, 1991
Building Act, 1984
Building Regulations, 1985
Caravan Sites and Control of Development Act, 1960
Clean Air Act 1993
Control of Pollution Act, 1974
Dangerous Wild Animals Act, 1976
Environment Act 1995
Environmental Protection Act 1990
Food Act 1984
Food Safety Act 1990
Health and Safety at Work etc. Act, 1974
Housing Act, 1985
Local Government and Housing Act 1989
Local Government (Miscellaneous provisions) Act, 1976 & 1982
Offices, Shops and Railway Premises Act, 1963
Pet Animals Act, 1951
Prevention of Damage by Pests Act, 1949
Public Health (Control of Disease) Act, 1984
Public Health Act, 1936, 1961
Refuse Disposal (Amenity) Act, 1978
Riding Establishments Act, 1964 1970
Shops Act, 1950
Slaughter of Poultry, 1967
Slaughterhouses Act, 1974
Zoo Licensing Act, 1981

**Form 3.1** General authorisation

purposes of an Act; to confuse the two is to invite disaster. An officer taking action when he or she is not authorised to do so will at the least prove to be embarrassing and at worst may leave the officer open to a legal action.

### 3.2.1.2 Authorisation for the Health and Safety at Work etc. Act 1974

Section 19 of the Act provides that every enforcing authority may appoint as inspectors persons having suitable qualifications as it thinks necessary. Section 19(2) makes it clear that every appointment of a person as an inspector must be made by an instrument in writing specifying the powers conferred. Under this Act such an inspector is only entitled to use those powers identified by the enforcing authority in the authorising document. It is now normal for there to be a separate document authorising an officer for the purposes of the Act (and other relevant statutory provisions made thereunder). Again an example of the wording of such a document is reproduced.

---

**Authorisation under the Health and Safety at Work etc Act. 1974**

**The Anytown City Council as the enforcing authority for the above Act and under the provisions of Section 19(1) of that Act hereby appoints .............................as an inspector for the purposes of that Act. As an inspector so appointed he/she is entitled to exercise all the powers of an inspector contained in Sections 20, 21, 22 and 25 of the Health and Safety at Work Act etc. 1974 and to exercise all the powers of an inspector contained in all the other relevant statutory provisions and to institute proceedings in England and Wales pursuant to Section 38 of the Health and Safety at Work etc. Act 1974**
**Dated this ...............day of ....................... 19...**

**Signed for and on behalf of Anytown City Council........................**

---

Form 3.2 Authorisation under the Health and Safety at Work etc. Act 1974

### 3.3 FEATURES OF THE POWER OF ENTRY

The schedule to the specimen authorisation makes it clear how extensive the powers of entry granted to an Environmental Health Officer are. Because of the range of provisions it is not surprising that there is no single, uniform procedure for gaining entry to premises. Each one of the statutory provisions displays common features, yet there is not sufficient uniformity to permit the Environmental Health Officer to assume that familiarity with one provides an adequate knowledge of all. Each is required to be understood in its own right. Each of the different statutory provisions permitting entry may display one or more of the characteristics shown in the following:

1. It permits entry by an authorised officer of a relevant authority
2. It requires the authorisation to be in writing
3. It permits entry by someone other than an authorised officer, e.g. a vet
4. It provides for any entry by an authorised oficer to be accompanied
5. It requires any authorised officer or other person using the power of entry to produce some form of written authorisation
6. It restricts entry for only specified reasons
7. It permits entry for general purposes of the Act or regulations
8. It requires prior notice (usually 24 hours) of intention to enter to be given to the occupier of the premises
9. It permits entry onto the premises without a warrant at any time
10. It permits entry onto the premises without a warrant at any reasonable time
11. It permits entry onto the premises without a warrant only between specified times
12. It grants a power to obtain a warrant for entry
13. It requires there to be prior notice to the occupier before a warrant can be obtained
14. It provides for the obtaining of a warrant in situations of emergency
15. It provides for the obtaining of a warrant in situations where prior notice would defeat the purpose of entry
16. It provides for the obtaining of a warrant where the situation is one of urgency
17. It provides for the obtaining of a warrant in situations where admission to the premises has been refused
18. It provides for the obtaining of a warrant in situations where refusal of admission to the premises is apprehended
19. It provides for the obtaining of a warrant where the premises are unoccupied
20. It provides for the obtaining of a warrant where the occupier is temporarily absent
21. It specifies that a warrant shall be of fixed duration
22. It specifies that a warrant shall not be of fixed duration
23. It permits the use of force to gain entry
24. It makes provision for entry outside the relevant authority's area
25. It permits entry to a dwellinghouse
26. It creates the offence of obstruction.

With such a multiplicity of features it is up to the Environmental Health Officer to ensure that he or she knows which of these is a feature of the provisions they are seeking to use. The assumption of similarity between statutory provisions could prove costly. If we consider for a moment two of the most common statutory provisions encountered by the Environmental Health Officer, the Environmental Protection Act 1990 (as amended by the Environment Act 1995) and the Food Safety Act 1990. Section 108 of the Environment Act 1995 (which replaced s.17 of the Environmental Protection Act 1990) confers upon inspectors an extensive list of investigative powers backed by criminal penalties for non-compliance. Subsection 4 and Schedule 18 provide that an authorised person may, on production of his designation and authority, exercise any of the powers in paragraphs a to m for the purposes of enabling the enforcing agency to carry out any assessment or prepare any report which the agency is required to carry out or prepare.

The powers of an inspector to enter are:

(a)  at any reasonable time (or, in an emergency at any time and if need be by force) to enter any premises which he has reason to believe it is necessary for him to enter;
(b)  on entering any premises by virtue of paragraph (a) above to take with him:
  (i)  any person duly authorised by the enforcing authority and, if the authorised person has reasonable cause to apprehend any serious obstruction in the execution of his duty, a constable and
  (ii) any equipment or materials required for any purpose for which the power of entry is being exercised.

Any authorised person must produce evidence of his designation and other authority before he exercises a power conferred by s.108 (Sch.18 para. 3).

Secion 108(7) provides that, except in an emergency, where an authorised person proposes to enter any premises and either:

(a)  entry has been refused and he apprehends, on reasonable grounds, that the use of force may be necesssary to effect entry or
(b)  he apprehends on reasonable grounds that entry is likely to be refused and that the use of force may be neccessary to effect entry, then any entry onto those premises under s.108 can only be effected under the authority of a warrant by virtue of Schedule 18 to the Act.

Schedule 18 of the Act states that if it is shown to the satisfaction of a justice of the peace on sworn information in writing:

1.  that there are reasonable grounds for the exercise in relation to any premises of a relevant power, and

2.   that:
    (a) the exercise of the power in relation to the premises has been refused
    (b) that such a refusal is reasonably apprehended
    (c) that the premises are unoccupied
    (d) that the occupier is temporarily absent from the premises and that the case is one of urgency or
    (e) that an application for admission would defeat the object of the proposed entry,

the justice may by warrant authorise an enforcing authority to designate a person who shall be authorised to exercise the power conferred by s.108 in relation to those premises, in accordance with the warrant, and, if need be, by force. Such a warrant will remain in force until the purposes for which it was issued have been fulfilled. Section 110 of the Act provides that the intentional obstruction of an authorised person in the exercise or performance of his powers or duties is an offence.

Thus in summary the features of the power of entry here are:

1.  It allows entry by an authorised person
2.  It requires any officer using the power of entry to produce evidence of his designation and other authority
3.  It permits entry for the purposes of the discharge of the functions of the enforcing authority
4.  It permits entry without a warrant at any reasonable time
5.  It permits entry without a warrant at any time (if certain conditions are met)
6.  It permits entry by force without a warrant (if certain conditions are met)
7.  It provides for the obtaining of a warrant
8.  It provides for the obtaining of a warrant in situations where admission to the premises has been refused, or refusal is apprehended
9.  It provides for the obtaining of a warrant where the premises are unoccupied or the occupier temporarily absent
10. It provides for the obtaining of a warrant where prior application for admission would defeat the purpose of entry
11. A warrant remains in force until the purposes for which it was issued have been fulfilled
12. It provides for any entry to be accompanied
13. It creates the offence of obstruction.

In contrast s.32 of the Food Safety Act 1990 provides that an authorised officer of an enforcement authority shall, on producing, if so required, some duly authenticated document showing his authority, have a right at all reasonable hours:

(a) to enter any premises within the authority's area for the purpose of ascertaining whether there is or has been on the premises any contravention of the provisions of the Act, or of regulations or orders made under it; and

(b) to enter any business premises, whether within or outside the authority's area for the purpose of ascertaining whether there is on the premises any evidence of any contravention within that area of any of such provisions; and

(c) in the case of an authorised officer of a food authority, to enter any premises for the purpose of the performance by the authority of their functions under this Act.

The section goes on to state that admission to any premises used only as a private dwellinghouse is not to be demanded as of right unless 24 hours' notice of the intended entry has been given to the occupier. This Act provides for the obtaining of a warrant and in subsection (2) states that if a justice of the peace, on sworn information in writing, is satisfied that there is a reasonable ground for entry into any premises and either:

(a) that admission to the premises has been refused, or a refusal is apprehended, and that notice of the intention to apply for a warrant has been given to the occupier; or

(b) that an application for admission, or the giving of such a notice, would defeat the object of the entry, or that the case is one of urgency, or that the premises are unoccupied or the occupier temporarily absent,

then the justice may by a warrant signed by him authorise an officer to enter the premises, if need be by reasonable force.

Subsection (4) stipulates that an authorised officer entering any premises by virtue of s.32 or of a warrant issued under it may take with him such other persons as he considers necessary and on leaving any unoccupied premises which he has entered by virtue of such a warrant he must leave them as effectively secured against unauthorised entry as he found them.

Every warrant granted under this section continues in force for a period of one month only (subsection (3)).

Section 33(1)(a) of the Act makes intentional obstruction of an authorised officer an offence.

The features of the power of entry under the Food Safety Act 1990 are:

1. It allows entry by an authorised officer
2. It requires any officer using the power of entry, if required, to produce written authorisation
3. It makes provision for entry outside the relevant authority's area

4. It permits entry for general purposes of the Act or regulations.
5. It permits entry to a dwellinghouse
6. It permits entry without a warrant at any reasonable time
7. It provides for any entry to be accompanied
8. It grants power to obtain a warrant
9. It requires there to be prior notice before a warrant can be obtained
10. It provides for the obtaining of a warrant in situations of emergency, or where prior notice would defeat the purpose of entry
11. It provides for the obtaining of a warrant in situations where admission to the premises has been refused, or refusal is apprehended
12. It provides for the obtaining of a warrant where the situation is one of urgency.
13. It provides for the obtaining of a warrant where the premises are unoccupied or the occupier temporarily absent
14. It permits the use of force
15. It specifies that a warrant shall be of fixed duration
16. It creates the offence of obstruction.

Though superficially similar in that they grant a power of entry the two Acts, the Environment Act 1995 and the Food Safety Act 1990, have differing requirements, provisions and limitations when dealing with a power of entry. These differences and those of the other statutory provisions must be noted and accommodated by all EHOs intending to use any formal power of entry. Table 3.1 summarises some of the main features of some commonly encountered provisions and shows how they may vary.

### 3.3.1 REASONABLE GROUNDS

The availability of statutory powers of entry is commonly limited to those situations where an officer feels there are 'reasonable grounds' for exercising the power or where the Justice, when issuing the warrant, is satisfied that the officer applying for the warrant has reasonable grounds to seek entry. Such words serve to impose a limit on the exercise of what might otherwise be an arbitrary power. Quite what are reasonable grounds will vary from case to case and it is not possible here to produce a definitive list of what will, in all situations, constitute such grounds. This will, in the final analysis, be a question of fact to be tried on evidence and must remain as part of an officer's professional judgement. What is clear however is that the test will be an objective one. In *Nakkuda Ali v Jayaratne* [1951] the House of Lords considered the meaning of the phrase 'reasonable grounds'. Lord Radcliffe held that the use of the words 'reasonable grounds' imposed a condition that there must in fact exist such reasonable grounds, known to an officer, before he can validly exercise the condi-

**Table 3.1** Features of power of entry of some statutory provisions

| | ABE 63 s.2 | BA 84 s.95 | CDCDA 60 s.26 | COP74 s.91 | EA95 s.108/Sch 18 | FSA90 s.32 | HSAWA 74 | EPA90 s.81 Sch 3 | LGMPA82 s.17 |
|---|---|---|---|---|---|---|---|---|---|
| Entry by an authorised officer | ✓ | ✓ | ✓ | ✓ | | ✓ | ✓ | ✓ | ✓ |
| Authorisation to be in writing | ✓ | ✓ | ✓ | ✓ | | ✓ | ✓ | ✓ | ✓ |
| Entry by other than authorised officer | ✓ | ✓ | ✓ | ✓ | ✓ | ✓ | ✓ | ✓ | ✓ |
| Production of written authorisation | ✓ | ✓ | ✓ | ✓ | ✓ | ✓ | ✓ | ✓ | |
| Prior notice of intended entry | ✓ | ✓ | ✓ | | ✓☉ | ✓* | | ✓* | |
| Entry without a warrant at any reasonable time | ✓ | | ✓ | | ✓ | ✓ | ✓ | ✓ | |
| Entry without a warrant at any time | | ✓ | ✓ | ✓ | ✓ | ✓ | ✓ | ✓ | ✓ |
| Power to obtain a warrant | | ✓ | ✓ | ✓ | ✓ | ✓ | | ✓ | ✓ |
| Prior notice of warrant | | ✓ | ✓ | ✓ | ✓ | ✓ | ✓ | ✓ | ✓ |
| Warrant in situations of refusal, emergency, etc. | | ✓ | | ✓ | ✓ | ✓ | | ✓ | |
| Warrant where premises are unoccupied | | ✓ | ✓ | ✓ | ✓ | ✓ | | ✓ | |
| Permits the use of force | ✓ | ✓ | ✓ | ✓ | ✓ | ✓ | ✓ | ✓ | ✓ |
| Restricts entry for only specified purposes | | | | | | ✓ | | | |
| Permits entry for general purposes | | ✓ | ✓ | ✓ | ✓ | ✓ | | ✓ | |
| Provision for entry in another LA | | | | | ✓ | ✓ | | | |
| Permits entry to a dwellinghouse | | ✓ | | ✓ | ✓ | ✓ | ✓* | ✓ | ✓ |
| Entry to be accompanied | | ✓ | ✓ | | | ✓ | ✓* | | |
| Warrant of fixed duration | | ✓ | ✓ | ✓ | ✓ | ✓ | ✓ | ✓ | ✓ |
| Warrant not of fixed duration | | ✓ | ✓ | ✓ | ✓ | ✓ | | ✓ | ✓ |
| Creates the offence of obstruction | ✓ | ✓ | ✓ | ✓ | ✓ | ✓ | ✓ | ✓ | ✓ |

✓ Feature present  * For a dwellinghouse  ✱ As residual general power  ☉ For residential premises

tional power in question. The issue is thus not, did the inspector believe there to be reasonable grounds but, when looked at objectively, were there reasonable grounds for so believing. An EHO must ask himself the question: 'would a reasonable person, possessed of the knowledge I have, believe there to be reasonable grounds for exercising a power of entry' and not simply 'do I believe there are reasonable grounds'.

There is limited precedent in this area, however judicial guidance may be obtained by an examination of related cases involving police officers and similar public officials. In *Dallison v Caffrey* [1965], which concerned a case of alleged false imprisonment and malicious prosecution by Detective Constable Caffrey, Lord Diplock stated (at 371e):

> The test whether there was a reasonable and probable cause for the arrest or prosecution is an objective one, namely whether a reasonable man, assumed to know the law and possessed of the information possessed by the defendant would believe that there was reasonable and probable cause.

The credibility in terms of the quantity and, particularly, quality of information which is required to establish 'reasonable grounds' necessarily must be of a lesser degree than that which would be required to institute proceedings or take other statutory action, for example service of a notice. In fact in many instances the powers of entry are exercisable as a prelude to ascertaining if a local authority should exercise any of its statutory powers. Such is the case for example in s.395 of the Housing Act 1985. Nevertheless Environmental Health Officers must ensure that they can satisfy the central objective test. This means that all officers should record clearly those facts which lead them to conclude that they have reasonable grounds for their belief. In so doing there is a twofold benefit. Firstly it ensures the officer has approached the issue correctly and satisfactorily addressed this threshold test for entry. Secondly it will ensure that, if subsequently questioned, on application for a warrant, during proceedings etc., the officer has clear and defensible reasons for coming to the conclusion he did.

### 3.3.2 REASONABLE TIMES

Not only are EHOs often required to exercise their powers of entry having regard to 'reasonable grounds' but also they are equally often limited to entering only at 'reasonable times'. Once again what will constitute reasonable times will be a question of law. It would, for example, generally be reasonable to enter a 24 hour cafe at 2 am (this being during business hours). However it would, generally, not be reasonable to enter a corner greengrocers at such times, though in certain circumstances entry to those premises at that time may be justifiable.

It is justifiable because implicit in the concept of reasonable time must be a consideration of the purpose of entry. Entry at 2 am to a corner shop for a routine visit is hardly likely to be reasonable. But were it to be suspected that, at that time, unfit food was being prepared for distribution for human consumption then, as this was only likely to be detected at 2 am, such a time, having regard to purpose, would be reasonable.

In the case of *Small v Bickley* (1875) it was held that entry on a Sunday afternoon, when the shop in question was closed, was not a reasonable time. The appellant, a butcher, was at his home half a mile from his shop. On the Sunday afternoon in question he was requested to go himself, or send someone with a key, to admit an inspector to his shop. This was required to allow examination of some meat on the premises. The butcher refused and was subsequently convicted of obstruction. On appeal it was held that although Sunday afternoon might in some circumstances be a reasonable time for the examination of meat, in this case the appellant had committed no offence. This finding is consistent with the overall approach of the courts in this area. In *Davies v Winstanly* (1931), a police officer entered a shop and wished to inspect the drugs register. The manager of the shop in question was at lunch and the register was locked in a cupboard. The Drugs (Consolidation) Regulations 1928 required the register to be available for inspection at all times. Mr Justice Avory in his judgment gave the following interpretation on the construction of the words 'at all times'.

> These words cannot be read literally to mean at all times of the day or night nor every day of the year ... in my opinion a reasonable construction was that ... the words may mean 'at all reasonable times.

The idea of 'reasonable time' ought not, in practice, to cause the Environmental Health Officer too many problems. What is reasonable will depend on:

- the nature of the premises to be entered
- the nature of the enquiry, investigation etc. which requires entry
- the time of day
- if an offence is involved or suspected, how serious the offence is: is there a threat to public safety, health etc?
- the practicability of entering at another time.

Entry onto domestic premises at 2 am would not normally appear to be reasonable. If however it was for the purposes of seizing stereo equipment causing a noise nuisance (see s.81(3) of the Environmental

Protection Act 1990) then it may be. In any event it is good professional practice to ensure that the requirement is seen to have been overtly acknowledged in any records the officer may keep. This may prove vital should the decision to enter prove to be challenged at a later date.

### 3.3.3 NOTICE OF ENTRY

It is again a common feature, before the exercise of a power of entry, to require there to be prior notice of the entry given to named persons. Section 32 of the Food Safety Act 1990 stipulates that admissions to any premises used only as a private dwelling-house shall not be demanded as of right unless 24 hours notice of the intended entry has been given to the occupier. Similarly s.91 of the Control of Pollution Act 1974 requires there to have been seven days notice of intended entry before a justice of the peace can issue a warrant. A failure to comply with the requirement of notice will render any dependent entry invalid.

A number of provisions requiring notice of entry require that the notice be given 'Not less than' a given numbers of days before the proposed entry. Such requirements raise questions as to which days are included for the purpose of calculating periods of notice. The first point to note, made by Lord Mansfield in *Pugh v Duke of Leeds,* is that date does not mean the hour or the minute but the day of delivery and in law there is no fraction of the day. Thus the general rule of the law is that in the computation of time, fractions of a day are not reckoned. The importance of correctly observing all formalities in connection with exercising a power of entry cannot be overemphasised and it is vital that time periods are correctly interpreted. The case of *R v Turner* [1910] concerned the relevant time periods of notice required when inserting a charge on an indictment. Channel J here held that 'not less than seven days notice' must be given a meaning of seven clear days notice.

In addition to notice of entry a number of other notices stipulate their requirements in terms of 'not less than X days'. A common notice served by an EHO is a request for information under s.16 of the Local Government (Miscellaneous Provisions) Act 1976. This provision stipulates that the recipient shall have a period to respond which is not less than 14 days beginning with the day on which the notice is served. How then should that period be calculated? In *Re Railway Sleepers Supply Co.,* Chitty J said that an interval of not less than 14 days is equivalent to saying that 14 days must intervene or elapse between the two dates. The test cited in *Webb v Fairmaner* was to reduce the time to one day. If the statute had said an interval of not less than one day, then if the first meeting is held on the 1st of January the second could not properly be held until the 3rd. One day must intervene and thus adding 13 more days to make up the 14 would mean that the second meeting could not be held before the 16th of January. On the same points see also *Chambers v Smith* (1843).

Where a statute specifies a period it should always be taken to mean clear days. This will not include the day the notice is received and it does not include the day on which the power of entry is exercised.

Any prior notice of entry must be specific as to its purpose. In the case of *Stroud v Bradbury* [1952] there was an attempted entry under the provisions of the Public Health Act 1936. The owner of a bungalow, having lost an earlier appeal, failed to carry out works required in accordance with a notice served under s.39 of the Public Health Act 1936 (see now s. 59 Building Act 1984). In consequence the local authority wrote to her on 17 February 1951 in which they said:

> this matter has now been held over for a considerable time since the appeal was dismissed by the quarter sessions committee, and as you have not complied with the council's notice it is intended to proceed under S.290(6) of the Public Health Act 1936.

Section 290(6) provided that if the person required by the notice to execute works failed to do so in the time indicated, the local authority could themselves execute the works and recover from that person the expenses reasonably incurred in so doing. And, without prejudice to their right to exercise that power, the recipient of the notice was liable to a fine. To be entitled to enter the premises the council had to observe the provisions of the Public Health Act 1936, s.287. Section 287 of that Act stated that:

(1) Any authorised officer of a Council shall, on producing a duly authenticated document if required, have a right to enter any premises at all reasonable hours.

(a) For the purposes of investigating contraventions of the Act.

(b) To ascertain whether the local authority should take any action under the Act.

(c) Generally for the performance of its duties under the Act.

Admission to premises other than a factory, workshop or workplace may not be demanded as of right unless 24 hours notice has been given. Section 283 of the Act required such notice to be in writing. This notice of intended entry was sent on 17 February 1951. On 1 June 1951 a sanitary inspector, accompanied by a builder, attended at the premises. It was the inspector's intention to ensure that the relevant works were carried out. He was at that time refused entry by Stroud, the occupier's husband, and threatened with assault if any attempt at entry was made.

Stroud was subsequently convicted of wilfully obstructing the inspector contrary to s.288 of the Public Health Act 1936 and was fined. He appealed and the matter ultimately reached the Divisional Court. Here it was held that the refusal by Stroud was lawful. The power of entry was given, not by s.290 of the Public Health Act 1936 but by s.287(1). Lord Goddard stated that the contention that an expression of intention on 17 February, to proceed under s.290(6) of the Public Health Act 1936 on 1 June, can possibly be a 24 hours notice of the intended entry is quite untenable.

Section 287(1) required 24 hours notice of intention to enter and the letter sent in February did not meet that requirement. It was not sufficiently specific in its wording such as to constitute adequate notice under s.287(1), and therefore the inspector had no right to enter. The appellant was entitled to order him off and prevent him from entering.

Any notice to enter should identify the relevant authority and it should be specific with regard to the relevant statutory provision concerned. Because of the often obligatory nature of such requirements a failure to observe the requirements of such a notice in scrupulous detail may result in there being no opportunity to institute proceedings for an offence of obstruction should entry be refused.

To avoid such a state of affairs notice of entry should:

1. Conform with any statutorily prescribed format. If there exist prescribed forms use them; for example Form 9 of the Housing (Prescribed Forms) (No 2) Regulations 1990 provides for a form of notice for use under the Housing Act 1985.
2. Ensure that it is given having regard to the correct timescale stipulated within the relevant statute.
3. Identify the authority that is the source of the notice, including postal address and telephone numbers.
4. Clearly identify the relevant statutory provision under which the notice is given.
5. Identify clearly and unambiguously the identity of the person to whom the notice is addressed.
6. Identify clearly and unambiguously the premises which are the subject of the notice.
7. Identify the reasons for the proposed entry.
8. Identify the officer proposing to exercise the power of entry.
9. State clearly the day, date, and with reasonable accuracy the time of the proposed entry. This ought to be sufficiently accurate to allow for foreseeable delays but not such as to allow the Environmental Health Officer to exercise the entry at some time during their day's duties.
10. The notice of intention should be signed, the author being clearly identified by name and designation.
11. The document should be dated and this should ensure that the period of time required for notice is clearly accommodated.
12. Have appropriate accompanying explanatory notes, particularly explaining what the recipient must do if the time proposed is not convenient and what the consequences, if any, of failing to allow entry may be.

**Anytown City Council**
**Town Hall**
**Anytown**
**Telephone : 01111 123456**

### NOTICE BEFORE EXERCISING POWER OF ENTRY
### HOUSING ACT 1985
### Section 197

To   Mr James Richard Smith
of   1 Green Street,
     Anytown
     AN1 1AA

You are the owner of the premises known as 22 Green Street, Anytown, AN1 1AA ('the premises').

Under section 197 of the Housing Act 1985, I Freda Brown being a person authorised in writing by the Anytown City Council, intend, on Wednesday the 11th December 1996, between the hours of 10.30 am and 11.00 am to enter the premises for the purpose of survey and examination.

It appears to the Council that survey and examination is necessary in order to decide whether any powers under Part VI of the Housing Act 1985 (repair notices) should be exercised in respect of the premises.

*Signature* ..................................................................)

Description
of the
person
authorised
to enter
Dated..............................

Form 3.3 Notice before exercising Power of Entry

**Anytown City Council**
**Town Hall**
**Anytown**
**Telephone : 01111 123456**

### NOTICE OF INTENTION TO EXERCISE A POWER OF ENTRY
### PUBLIC HEALTH ACT 1936
### Section 287

To    Mr James Richard Smith
of    1 Green Street,
       Anytown
       AN1 1AA

Whereas you are the owner the premises known as 22 Green Street, Anytown, AN1 1AA (hereinafter the 'premises').

It is the opinion of the Anytown City Council (hereinafter the Council ) that it is necessary for an authorised officer of the council to enter the premises for the purposes of the Public Health Act 1936.

Note then that, under section 287 of the Public Health Act 1936, I Freda Brown being a person authorised in writing by the Council, hereby give notice that it is my intention on Wednesday the 11th December 1996, between the hours of 10.30 am and 11.00 am to enter the premises.

Signature ................................................................

Description
of the
person
authorised
to enter
Dated...............................

Form 3.4  Notice of intention to exercise a Power of Entry

## 3.4 WARRANTS

### 3.4.1 CONTENT

Though EHOs generally can and will often have to apply for warrants of various types it may be an individually infrequent activity. There can therefore often be something of a mystery surrounding such things. There may be a belief that such a document must carry the Royal crest, be written on vellum in a copperplate hand, perhaps even need to be tied with red tape. None of these are necessary. A warrant in the circumstances discussed here is a working and workable legal document. The authority it carries is not derived from its form but from its content. It would be unwise to believe that a disappointing presentation in any way robbed a warrant of its authority. It gives a rare power which must be exercised properly and thus it is that any purported exercise of these powers will, in the words of Lord Wilberforce in *R v IRC ex p. Rossminster Ltd* [1980], be supervised critically, even jealously by the courts. A warrant was said by Lord Wilberforce in *Rossminster* to be a document issued by a person in authority under power conferred in that behalf authorising the doing of an act which would otherwise be illegal. Here it makes the entry, even forcible entry, legal. Generally speaking a warrant does not have to take a particular form but must generally have a particular content. The House of Lords in *Rossminster* said that a warrant should satisfy the person affected that the power to issue exists, therefore the warrant should contain reference to that power. It is also wise to add a statement of satisfaction on the part of the judicial authority as to the matters on which he must be satisfied. Lord Diplock considered that a warrant should further contain details of the authority under which it is issued (the relevant statutory provision), the name of the person to whom it is addressed (the authorised officer) and the address of the premises to be entered. Providing these criteria are met and providing other formalities have been followed a legally enforceable warrant will then exist.

### 3.4.2 OBTAINING A WARRANT

For the Environmental Health Officer a warrant is usually obtained by applying to a justice of the peace. This is accomplished by presenting information on oath to the justice regarding the proposed entry. The information should be in writing, it should closely follow the wording of the statutory provisions under which it is sought and it should set out details relating to the application. Any officer applying for a warrant is also normally expected to prepare the (unsigned) warrant in advance and present it to the justice of the peace at the same time as the information.

Authorities may have, indeed should have, set procedures for the obtaining of warrants and generally applications will follow an established procedure. On application for a warrant an officer will normally present himself at an agreed time at the magistrates court having jurisdiction for the premises in question. In the ordinary course of events this should be in the morning before the routine business of the court has commenced. Once sitting the magistrates would not expect to be distracted by an unexpected application for a warrant and any applicant may find it difficult to get an unscheduled place on an already busy court list.

An officer applying for a warrant would speak first to one of the court clerks and explain the purpose of his visit. The clerk will normally wish to have details of the applicant, the application and see copies of both the information and the proposed warrant. The clerk will wish to ensure that:

1. there is territorial jurisdiction to issue the warrant
2. the applicant is qualified to apply for the warrant
3. the information and warrant conform to the statutory provisions.

The officer should thus ensure that he takes with him all instruments of appointment relevant to the application. To have got this far only to realise that you cannot prove you are an authorised officer is a little embarrassing. The actual application before the justice may take place in open court or may occur in one of the other rooms in the court building. An officer will be required to swear an oath or affirm the truth of his information. He may then be questioned by the justice who has a duty to satisfy himself that in all the circumstances it is right to issue the warrant. Thus the officer may expect to be questioned about:

(a) who he is
(b) his authority to apply for the warrant
(c) confirmation of the address of the premises to be entered
(d) confirmation that all appropriate procedures have been followed
(e) the reliability of the officer's information
(f) enquiries he has made and the information he has obtained.

Particularly the justice will seek information on, and assurances that, the appropriate notices of intention to seek a warrant were given or, as the case may be, that admission has been refused or refusal is apprehended or that the case is one of urgency etc.

An officer may and often will be questioned about the circumstances surrounding the application in order that the justice may be satisfied that there are sufficient grounds to merit the issue of a warrant. The officer must be open and frank. It may be necessary for the officer to refer to papers to confirm dates of events. The taking of a case file can thus prove to be very useful, though there is no requirement for the magistrate to have sight of the contents of such a file.

Magistrates are made aware that the issuing of a warrant is a serious step which will interfere with a person's liberty and right to privacy; it is for these reasons that, before the issuing of a warrant, a justice will wish to satisfy himself that it is right to do so in all the circumstances.

It would be unwise to assume that the obtaining of a warrant is merely an administrative inconvenience as a prelude to entry. Justices are mindful that the wrongful issue of a warrant could be personally damaging to the individual whose premises are to be entered and make them, the justices, personally liable in damages if they exceed their powers. There is therefore no automatic right to a warrant and the decision whether or not to issue a warrant is entirely at the justice's discretion. It is therefore up to the applicant to convince the magistrate of the correctness of his application. When the enquiry is concluded the officer should sign the information and the justice will sign both the information and the warrant. The information will be retained for court records, and it is important that the officer ensure he has a copy for his own records.

A specimen application for a warrant is reproduced.

## APPLICATION FOR WARRANT TO ENTER PREMISES
## ENVIRONMENTAL PROTECTION ACT 1990:
## SECTION 81/SCHEDULE 3

This is the information and application on oath of Freda Brown a duly authorised officer of Anytown City Council.

Who states that for the purposes of paragraph 2 of Schedule 3 to the Environmental Protection Act 1990 she has a right at any reasonable time to enter the premises known as 1 Green Street Anytown AN1 1AA [the premises] occupied by James Richard Smith and that it is necessary that a duly authorised officer of the said council should enter the said premises.

And that, after giving notice of entry on the 10th December 1996, admission to the said premises was refused to the said Freda Brown on 19th December 1996 and that there are reasonable grounds for entry to the premises.

Signed [Applicant].........................................

Taken and Sworn before me this ....................day of...................., 19......

Justice of the Peace

**Form 3.5** Application for warrant to enter premises

## WARRANT TO ENTER PREMISES
## ENVIRONMENTAL PROTECTION ACT 1990:
## SECTION 81/SCHEDULE 3

TO: Freda Brown

Information on oath has been laid today by Freda Brown Environmental Health Officer of Anytown City Council that for the purposes of paragraph 2 of Schedule 3 to the Environmental Protection Act 1990 she has a right at any reasonable time to enter the premises known as 1 Green Street Anytown AN1 1AA occupied by James Richard Smith

And for the purpose of exercising that right, she applies for a Warrant to be issued for the reason that on 19th December 1996 admission to the said premises was refused, a notice pursuant to Schedule 3 of the Environmental Protection Act 1990 having been previously served on the said James Richard Smith on the 10th December 1996 informing him of her intention to enter the premises on the 19th December 1996.

And that there is reasonable ground for entry to the said premises for the purpose of the performance by the Council of their functions under Part III of the Act and specifically to carry out works to abate a statutory nuisance.

I the undersigned, a Justice of the Peace for the Anytown Petty Sessional Division, being satisfied of the truth of the said information and of there being reasonable grounds for entry into the said premises, by this Warrant authorise the said Freda Brown, taking with her such other persons as may be necessary, to enter the said premises, if need be by force, for the purpose aforesaid, pursuant to Schedule 3 of the Environmental Protection Act 1990.

Dated this ....................day of......................, 19......

COURT

STAMP

Justice of the Peace

**Form 3.6** Warrant to enter premises: Environmental Protection Act 1990

## WARRANT TO ENTER PREMISES
## PUBLIC HEALTH ACT 1936:
## SECTION 287

TO: Freda Brown

Information on oath has been laid today by Freda Brown Environmental Health Officer of Anytown City Council that for the purposes of section 287 Public Health Act 1936 she has a right at any reasonable time to enter the premises known as 1 Green Street Anytown BK1 1AA occupied by James Richard Smith

And for the purpose of exercising that right, she applies for a Warrant to be issued for the reason that on 19th December 1996 admission to the said premises was refused, a notice pursuant to section 287 Public Health Act 1936 having been previously served on the said James Richard Smith on the 10th December 1996 informing him of her intention to enter the premises on the 18th December 1996.

And that there is reasonable ground for entry to the said premises for the purpose of the performance by the Council of their functions under the Act.

I the undersigned, a Justice of the Peace for the Anytown Petty Sessional Division, being satisfied of the truth of the said information and of there being reasonable grounds for entry into the said premises, by this Warrant authorise the said Freda Brown, taking with her such other persons as may be necessary, to enter the said premises, if need be by force, for the purpose aforesaid, pursuant to section 287 of the Public Health Act 1936 .


Dated this ....................day of......................, 19......

COURT
STAMP

Justice of the Peace


**Form 3.7** Warrant to enter premises: Public Health Act 1936

### 3.4.3 EXECUTION AND EXPIRY

Unless the statute in question specifically restricts the time that the power of entry may be used then entry under a warrant may be at any time (*R v Adams* [1980], *R v IRC ex p. Rossminster Ltd* [1980]). Subject to any time limit imposed by the statute a warrant once issued remains in force until the purposes for which it was given have been satisfied. Once executed the warrant, even if not time expired, should be considered spent (*R v Adams* [1980]). The idea of repeated visits to premises under a single warrant is not satisfactory and should be avoided. It is therefore incumbent on any officer entering under a warrant to ensure that they carry out a full and thorough investigation at that time. Repeated applications for warrants to enter, attributable to inadequate professionalism on the part of the applicant, are unlikely to meet with success. At the conclusion of any inspection or examination any premises entered by force must be made as secure against unauthorised entry as they were at the time of entering.

## 3.5 OBSTRUCTION

### 3.5.1 PHYSICAL ELEMENT

Where a power to enter is given there exists the corresponding obligation to permit entry. Yet even with a warrant it is not uncommon for an officer to face opposition This resistance can take an number of forms ranging from unhelpfulness to physical obstruction. As a number of provisions provide for an offence of obstruction it is important that an Environmental Health Officer is familiar with what might constitute this offence.

Essentially the *actus reus* of obstruction may be taken to be any act which may make it more difficult for an officer to carry out his or her duties (*Lewis v Cox* [1984]). In *Hinchcliffe v Sheldon* [1955] it was found that the offence of wilful obstruction had been committed when an accused had made it more difficult for a police officer to carry out his duties. In this case a person accused under s.2 of the Prevention of Crimes Amendment Act 1885 had simply called out warnings to the occupier of premises the police wished to enter. From these authorities it is clear that obstruction need not involve the creation of any sort of physical barrier and that passive resistance may be enough, though not passive unhelpfulness (*R v Chief Constable of Devon and Cornwall ex p. CEGB* (1982)). A failure to carry out an act where there is no legal duty requiring it does not constitute an obstruction (*Swallow v L.C.C.* [1916]). In *Barge v British Gas Corporation* (1983) the court found that a refusal to hand over or copy the relevant document did not contravene section 28 of the Trade

Descriptions Act 1968, which required only that documents be produced. No offence had been committed by the refusal to comply with the wishes of a Trading Standards Officer. In giving access to the papers, those charged had met their legal duty. The section of the Act did not empower the inspector to make further demands. In *Green v DPP* (1991) it was held that advising someone not to say anything, when that person is being questioned by the police, did not amount to obstructing a constable in the exercise of his duty. Any person has the right to give advice to someone as to their right to refuse to answer the questions put by a police officer. It follows that this will extend to officers involved in the enforcement of other areas of the criminal law.

### 3.5.2 MENTAL ELEMENT

More recent legislation enforced by an Environmental Health Officer now frames the offence of obstruction with reference to intentional, rather than wilful, conduct. For example the Food Safety Act 1990, s.33 departed from the previous wording contained in the Food Act 1984. That Act made it clear that anyone who wilfully obstructed any person acting within the provisions of the Act was liable on conviction to a fine. The newer Act has replaced this with a requirement that the obstruction be 'intentional'. Though some of the more recent legislative updating in other areas has involved the removal of the wilful element of the offence and its replacement with intent there remain a number of older, though still valid, provisions which continue to frame the offence of obstruction by reference to wilful conduct, for example the Public Health Act 1936. Wilful obstruction has been taken to require deliberate and intentional acts.

In *Arrowsmith v Jenkins* [1963] the Court of Appeal found that that if a person without lawful authority or excuse intentionally, as opposed to accidentally, does something or omits to do something he is guilty of an offence. Violence is not an essential element of this offence; though where it is found to have been present it is likely to be conclusive of the conduct being wilful.

As has already been discussed, there is no general agreement in English law as to the meaning of intention nor has it been statutorily defined. In the case of *R v Mohon* [1975] intention was taken to mean a decision to bring about, insofar as it lies within the accused's power, a particular consequence, no matter whether the accused desired the consequence of his act or not. In *Lewis v Cox* [1984]) the court said that the test for wilfulness was whether the defendant intended his conduct to prevent the officers from carrying out their duties or to make it more difficult for them to do so. Effectively, having regard to the foregoing the two states of mind, intention and wilfulness, are equivalent terms.

### 3.5.3 FORCE

As has already been noted it is in certain instances permissible to gain entry to premises by the use of force. In *Swales v Cox* (1981), which concerned police entry onto premises to carry out an arrest, it was held that force is used if an officer applies any energy to an obstacle with a view to removing it. So where a door is ajar and energy is applied to make it open further, force is used. When a door is latched, the handle is turned and the door eased open, force is used, and if anyone opens any window or increases the opening in any window, or indeed dislodges the window by the application of any energy, then that is using using force to enter. 'Entry by force meant entry upon premises without an invitation to do so.' If this is so and if entry by force is only permissible with a warrant then any Environmental Health Officer entering without permission not only does so as a trespasser but does so with the unauthorised use of force.

In *Grove v Eastern Gas Board* [1952] it was conversely found that a power of entry conferred by statute authorised forcible entry, if it was necessary. In *Grove* the gas board sought entry to empty and inspect the gas meter and other appliances. Due notice was given to the occupier of the intended entry and a call was made on 27 June; however entry was not obtained. After further unsuccessful calls on 7, 12 and 26 July officials eventually removed a pane of glass near to the front door and obtained entry to the premises. On leaving they replaced the glass and left the premises no less secure than on entry. The occupier sued for trespass and claimed damages. The action was unsuccessful, Somerville LJ finding that: '... a power of entry conferred by a statute is, prima facie, at any rate, a power of forcible entry if necessary'.

The two apparently conflicting decisions, on the face of them, may cause practitioners a problem. Perhaps the intelligent approach to these two conflicting cases is to recognise that use of only minimal force, for example pushing a door open that is already ajar, is not something that ought to require judicial supervision. It is unlikely to be considered to be an application of force disproportionate to that which may be implied under the statutory power. On the other hand 'true forcible entry', i.e. breaking down of a door or window, picking of locks or removal of other obstacles, including people, which physically bar the way, does require the authority of the court and should correctly only be undertaken after a warrant to do so has been obtained. Whenever there is any doubt an officer must err on the side of caution and not enter without authority of a warrant. In all circumstances the force applied should always be the minimum necessary to attain the objective.

Therefore when the statutory power is silent as to the use of force it may be possible to use minimum force but only as a last resort. If there is an alternative procedure, for example to obtain a warrant, this must be followed. When the use of force is authorised the power may be exercised without the need to first establish the refusal of non-forcible entry. But in all cases any force applied must always be limited to that which is reasonable.

## 3.6 TRESPASS *AB INITIO*

If a person enters under authority of law but then proceeds to commit an act not justified under his power of entry he will become liable as a trespasser. Moreover if he abuses his authority in this way he becomes a trespasser *ab initio* and he is treated as if he was a trespasser from the time he entered, no matter how innocent his conduct was prior to the time of the abuse of authority. It is important to note that this only applies where entry is gained under authority of law. In fact this doctrine has been much eroded and is, as a matter of practice, now of limited importance. The situation pertaining now is that a partial abuse will not render everything done under it unlawful. In *Elias v Passmore* [1934] two police officers entered the premises of the National Unemployed Workers movement to arrest a person, for whom an arrest warrant had been issued. While they were on the premises they unlawfully seized documents. The court found that this did not render the officers trespassers *ab initio* with regard to the premises, though it did with regard to the documents seized.

## 3.7 INDIVIDUAL DETAILS OF POWERS OF ENTRY

Having looked at the provisions with regard to entry generally the next section gives detail on the specific powers of entry to be found in some of the most commonly encountered statutory provisions administered by EHOs.

### 3.7.1 ANIMAL BOARDING ESTABLISHMENTS ACT 1963

The power to enter animal boarding establishments is to be found in section 2(1) of the Animal Boarding Establishments Act 1963. A local authority may authorise in writing any of its officers or any veterinary surgeon or veterinary practitioner to inspect, to prevent the spread among animals of infectious or contagious diseases, any premises in their area licensed under the Act. Any such authorised person on producing his authority if so required may enter at all reasonable times such premises to inspect them and any animals on them for the purposes of ascertaining whether an offence has been or is being committed. There is by virtue of s.2(2) the offences of 'wilfully obstructing' or 'delaying' any person in the exercise of powers. Anyone convicted of such an offence is liable to a fine not exceeding level 1 on the standard scale (s.3(3)).

### 3.7.2 BREEDING OF DOGS ACT 1973

Powers of entry are to be found in section 2(1) of the Act.

A local authority may authorise in writing any of its officers or any veterinary surgeon or veterinary practitioner to inspect, to prevent the spread among animals of infectious or contagious diseases, any premises in their area licensed under the Act. Any such authorised person on producing his authority if so required may enter at all reasonable times such premises to inspect them and any animals on them for the purposes of ascertaining whether an offence has been or is being committed.

Section 2(2) of the Act makes it an offence to wilfully obstruct or delay any person in the exercise of powers of entry. Section 3(2) specifies that the penalty for wilful obstruction or delay shall be a fine not exceeding level 3 on the standard scale.

### 3.7.3 BREEDING OF DOGS ACT 1991

This Act provides a power of inspection under the 1973 Act to include premises which are not covered by a licence under that Act. Section 1 of the Act allows a justice of the peace, if satisfied, on the sworn information of an authorised officer or an authorised veterinary surgeon or practitioner, that there are reasonable grounds for suspecting that an offence has been or is being committed contrary to s.1(1) of the Breeding of Dogs Act 1973, to issue a warrant. Such a warrant may authorise entry by force and remains in force for a period of one month commencing on the date of issue. The entry must be exercised at reasonable times and anyone entering under such a warrant may take with them others. The power of entry here does not extend to entry to a private dwelling. Anyone intentionally obstructing or delaying entry is guilty of an offence and liable on summary conviction to a fine not exceeding level 3 on the standard scale.

3.7.4 BUILDING ACT 1984

Sections 95 and 96 detail the powers of entry.

### 3.7.4.1 Section 95(1)

Any authorised officer of a local authority shall, on producing if so required some duly authenticated document showing his authority, have a right to enter any premises at all reasonable hours:

(a) For the purposes of ascertaining whether there is, or has been, on or in connection with the premises any contravention of the provisions of this Act or of any building regulations that it is the duty of the local authority to enforce.
(b) For the purposes of ascertaining whether or not circumstances exist which would authorise or require the local authority to take any action or execute any work under the Act or building regulations.
(c) For the purposes of taking any action or executing any work authorised or required by this Act or building regulations.
(d) Generally for the purpose of the performance of the local authority's functions under the Act or building regulations.

Admission to premises other than a factory or workplace may not be demanded as of right unless 24 hours notice of the intended entry has been given to the occupier.

If it is shown to the satisfaction of a Justice of the Peace on sworn information in writing:

(a) that admission to any premises has been refused, or refusal is apprehended, or that the premises are unoccupied or that the occupier is temporarily absent, or that the case is one of urgency, or that an application for admission would defeat the object of entry; and
(b) there is reasonable ground for entry into the premises for the purposes stipulated,

the Justice may by warrant authorise the Council by any authorised officer to enter the premises, if need be by force.

A warrant should not be issued unless the Justice is satisfied that the occupier has been given notice of intention to apply for a warrant or the premises are unoccupied, or the occupier is temporarily absent, or the case is one of urgency, or that such notice would defeat the object of entry.

An authorised officer entering premises by virtue of section 95 or of a warrant issued under it may take with him such persons as may be necessary and on leaving any unoccupied premises must ensure they are effectively secured against trespassers as when he found them. A warrant issued under this section remains in force until the purpose for which the entry was necessary has been satisfied.

### 3.7.5 CARAVAN SITES AND CONTROL OF DEVELOPMENT ACT 1960

Section 26 of the Act provides the power of entry to an authorised officer of the local authority.

The entry is exercisable at all reasonable hours to an existing caravan site or site for which application is being made. If demanded by the occupier the officer must produce his authorisation.

Under the Act four reasons for entry are covered. These are for:

(1) the setting or altering of site conditions;
(2) ascertaining whether contraventions of the Act exist or have existed;
(3) ascertaining whether circumstances on the site justify the local authority to take action or undertake works;
(4) the purpose of taking action or undertaking works on the site provided for by the Act.

Normally 24 hours notice must be given. This may be waived in certain specified circumstances.

Section 26(3) allows the authorised officer to take with him such other persons as may be necessary.

On application to a justice of the peace a warrant may be issued if the justice is satisfied that:

(a) admission to the premises has been refused, or a refusal is apprehended, and that notice of the intention to apply for a warrant has been given to the occupier; or
(b) the occupier is absent and that the case is one of urgency;
(c) an application for admission would defeat the object of the entry.

Every warrant granted under this section shall continue in force until the purpose for which it was granted is satisfied. Section 26(5) stipulates that 'wilful obstruction of any person' (this is not restricted to just the authorised officer) acting in the execution of this section shall be liable on summary conviction to a fine not exceeding exceeding level 1 on the standard scale.

### 3.7.6 CLEAN AIR ACT 1993

Section 56 of this Act provides that any person authorised by a local authority may at any reasonable time enter any land or vessel and carry out such inspections as he considers appropriate. This power of entry does not apply to dwellings except in relation to adaptations in smoke control areas. Subsection 3 of the Act gives an officer the ability to obtain a warrant.

If a justice of the peace is satisfied on sworn information in writing that:

(a) admission to any land or vessel is refused, or refusal is apprehended, or the land or vessel is unoccupied, or the occupier is temporarily absent, or the case is one of emergency or that application would defeat the object of entry; and
(b) there is reasonable ground for entry upon the land or vessel,

then the justice may by warrant authorise that person to enter if need be by force.

A justice must not issue a warrant under this section unless satisfied:

(i) that admission to the land or vessel was sought after not less than seven days notice of intended entry has been served on the occupier; or
(ii) that admission was sought in an emergency and was refused by or on behalf of the occupier;
(iii) that the land or vessel is unoccupied; or
(iv) an application for admission would defeat the object of the entry.

A warrant issued under this section remains in force until the purpose for which the entry was necessary has been satisfied.

Section 57 requires that anyone entering under this Act must, if so required, produce evidence of his authority before entering and must on leaving leave unoccupied land as secure against entry as when he found it. It is an offence to wilfully obstruct anyone exercising powers conferred under s.56 or s.57 and such a person will be guilty of an offence and liable on summary conviction to a fine not exceeding level 3 on the standard scale.

### 3.7.7 CONTROL OF POLLUTION ACT 1974

Section 91 of the Act provides that any person authorised in writing by a relevant authority may at any reasonable time:

(a) enter upon any land or vessel for the purpose of –
(i) performing functions conferred on that person or the authority by virtue of the Act
(ii) determining whether, and if so in what manner, such a function should be performed
(iii) determining compliances with the Act

(b) carry out such inspections, measurements and tests on land or vessel, or of any articles, and take away samples of land or articles as he considers appropriate for such a purpose. Powers of entry are contained in section 91 of the Act.

If it is shown to the satisfaction of a justice of the peace on sworn information in writing that:

(a) admission to any land or vessel is refused, or refusal is apprehended, or they are unoccupied, or the occupier is temporarily absent, or the case is one of emergency or that application would defeat the object of entry; and
(b) there is reasonable ground for entry upon the land or vessel,

then the justice may by warrant authorise that person to enter if need be by force.

A justice must not issue a warrant unless satisfied:
(i) that admission to the land or vessel was sought after not less than seven days notice of intended entry has been served on the occupier; or
(ii) that admission was sought in an emergency and was refused by or on behalf of the occupier.
(iii) that the land or vessel is unoccupied; or
(iv) an application for admission would defeat the object of the entry.

A warrant issued under this section remains in force until the purpose for which the entry was necessary has been satisfied. Section 92 deals with provisions which are supplementary to s.91 and provides that:

(1) An authorised person must, if so required, produce evidence of his authority before entering land or vessels.
(2) He is authorised to take with him other persons and equipment.
(3) Admission to land or vessels used for residential purposes or admission to land and vessels generally with heavy equipment shall not take place without seven days notice.

There are exceptions to these to deal with situations where the matter is one of emergency (a case in which a person requiring entry has reasonable cause to believe that circumstances exist which are likely to endanger life or health and that immediate entry is necessary to verify the existence of those circumstances or to ascertain their cause or to effect a remedy) or where land or a vessel is unoccupied. Officers exercising a power of entry under these provisions are obliged to ensure that land or vessels are, on leaving, effectively secured. A person who wilfully obstructs another person in the exercise of these powers shall be guilty of an offence and liable to fine not exceeding level 3 on the standard scale.

### 3.7.8 DANGEROUS WILD ANIMALS ACT 1976

Powers of entry are contained in section 3, although it is surprising to note that no power of forceable entry appears to be provided for under the statute.

### 3.7.9 ENVIRONMENT ACT 1995

Section 108 of the Environment Act 1995 deals with the powers of enforcing authorities and those authorised by them. This replaces earlier provisions found in sections 16 to 18 of the Environmental Protection Act 1990. This section of the Act confers upon authorised persons an extensive list of powers. The powers are exercisable by an authorised person for the purpose of:

(a) determining whether any provision of the pollution control enactments in the case of that authority is being or has been complied with
(b) exercising or performing one or more of the pollution control functions of that authority, or
(c) determining whether and if so how such a function should be exercised or performed.

An authorised person may be given powers under the Act. Such an authorised person has the power to enter any premises at any reasonable time (or, in an emergency, at any time and if need be by force). On entering any premises he may take with him any person duly authorised by the enforcing authority and, if the authorised person has reasonable cause to apprehend any serious obstruction in the execution of his duty, a constable and any equipment or materials required for any purpose for which the power of entry is being exercised. Any authorised person must produce evidence of his designation and other authority before he exercises a power conferred by s.108 (Sch.18 para. 3).

The powers contained in s.108 allow the authorised person to make such examination and investigation as may in the circumstances be necessary. An authorised person may direct that any premises that he has power to enter or any part of them or any thing in them is to be left undisturbed for so long as is reasonably necessary for the purpose of any examination or investigation. An authorised person may take measurements, photographs, recordings and samples of any articles or substances found in or on the premises. He may also sample the air, water or land in or in the vicinity of the premises. In the case of any article or substance found in or on the premises which he has power to enter, being an article or substance which appears to him to have caused or to be likely to cause pollution of the environment or harm to human health, an authorised person may have it dismantled or subject to any process or test. The authorised person may also take possession of the article or substance and detain it for so long as is necessary to examine it, to ensure that it is not tampered with before examination of it is complete and to ensure that it will be available as evidence in any proceedings for an offence.

### 3.7.10 FOOD SAFETY ACT 1990

The extensive powers of an authorised officer are to be found in section 32.

An authorised officer of an enforcement authority shall, on producing if so required some duly authenticated document showing his authority, have a right at all reasonable hours to enter any premises within the authority's area for the purpose of ascertaining whether there is or has been on the premises any contravention of the provisions of this Act, or of regulations or orders made under it; and to enter any business premises (irrespective of whether they are within or outside the authority's area) for the purpose of ascertaining whether there is on those premises any evidence of any contravention within that area of any provisions of the Act, or of regulations or orders made under it. An authorised officer of a food authority is able to enter any premises for the purpose of the performance by the authority of their functions under the Act. Admission to premises which are used only as a private dwelling-house may not be demanded as of right unless there has been 24 hours notice of the intended entry given to the occupier.

If a justice of the peace on sworn information in writing is satisfied that there is reasonable ground for entry into any premises for the purposes mentioned and either:

(a) that admission to the premises has been refused, or a refusal is apprehended, and that notice of the intention to apply for a warrant has been given to the occupier; or

(b) that an application for admission or the giving of such a notice would defeat the object of the entry, or that the case is one of urgency, or that the premises are unoccupied or the occupier temporarily absent, the justice may by warrant signed by him authorise the authorised officer to enter the premises, if need be by reasonable force.

Every warrant granted under this section shall continue in force for a period of one month.

Any authorised officer entering any premises by virtue of section 32 or of a warrant is empowered to take with him any other persons as he considers necessary. When leaving any unoccupied premises which have been entered they must be left as effectively secured against unauthorised entry as when found.

Section 33 deals with the offence of obstruction and provides that any person who

(a)  intentionally obstructs any person acting in the execution of this Act;
(b)  without reasonable cause, fails to give to any person acting in the execution of this Act any assistance or information which that person may reasonably require of him for the performance of his functions under this Act, shall be guilty of an offence.

A person guilty of an offence under this section is liable to a fine not exceeding level 5 on the standard scale or to three months imprisonment. Exception to the power of entry is provided in the case of cowsheds or any other place containing an infected animal or declared to be an infected place under the Animal Health Act 1981, in which circumstances separate authorisation must be obtained from the relevant local authority.

### 3.7.11 HEALTH AND SAFETY AT WORK ETC. ACT 1974

The powers of an inspector authorised under this Act extend beyond simple entry onto premises and are to be found in section 20 of the Act.

The powers of an inspector are that he may at any reasonable time (or, in a situation which in his opinion is or may be dangerous, at any time) enter any premises which he has reason to believe it is necessary for him to enter. He may also take with him a constable if he has reasonable cause to apprehend any serious obstruction in the execution of his duty and on entering any premises he may take with him any other person duly authorised by his, that is the inspector's, enforcing authority. Any authorised officer when exercising this power of entry may take any equipment or materials required for any purpose for which the power of entry is being exercised and may make such examination and investigation as may in any circumstances be necessary. The inspector may, as regards any premises which he has power to enter, direct that those premises or any part of them, or anything therein, be left undisturbed for so long as is reasonably necessary for the purpose of any examination or investigation.

It is an offence for a person to:

(a) obstruct any investigation
(b) intentionally obstruct an inspector in his powers or duties.

Section 33 deals with offences and the penalty for obstruction of an officer is up to a maximum of level 5 on the standard scale. Officers should note the residual power contained in the final paragraph of s.20 which may be of use when seeking to gain entry.

### 3.7.12 HOUSING ACT 1985

There are a number of different sections dealing with power of entry within this Act (Table 3.1).

**Table 3.1** Different sections dealing with power of entry in the Housing Act 1985

| Part | Main Area | Section |
|------|-----------|---------|
| II | Provision of accommodation | 54 |
| VI | Repair Notices | 197 |
| VIII | Area Improvement | 260 |
| IX | Slum Clearance | 319 |
| X | Overcrowding | 340 |
| XI | HMO | 395 & 397 |
| XII | Common Lodging Houses | 411 |
| XVII | Compulsory Purchase | 600 |

### 3.7.12.1 Section 54

A person authorised by a local housing authority or the Secretary of State may, at any reasonable time, on giving 24 hours notice (the form of notice is prescribed) of his intention to the occupier, and to the owner if the owner is known, enter premises (not just houses) for the purpose of survey and examination. This power is exercisable where it appears to the authority or Secretary of State that survey or examination is necessary in order to determine whether any powers under Part II should be exercised in respect of the premises, or in the case of premises which the authority is authorised by Part II to purchase compulsorily.

Any authorisation for the purposes of this section must be in writing stating the particular purpose or purposes for which the entry is authorised and shall if so required be produced for inspection by the occupier or anyone acting on his behalf.

### 3.7.12.2 Section 197

A person authorised by the local housing authority or the Secretary of State may at any reasonable time, on giving seven days notice (see Form 9 of the Housing (Prescribed Forms) (No. 2) Regulations 1990) of his intention to the occupier and to the owner if the owner is known, enter premises for the purpose of survey and examination where it appears to the authority that survey or examination is necessary in order to determine whether any powers under this Part of the Act should be exercised in respect of the premises.

The officer's authorisation for the purposes of this section must be in writing and state the particular purpose or purposes for which the entry is authorised. If required, the authorisation must be produced for inspection by the occupier of premises or anyone acting on his behalf.

Section 198 creates the offence of intentionally obstructing anyone authorised under the provisions and any person who commits such an offence is liable on conviction to a fine not exceeding level 3 on the standard scale.

### 3.7.12.3 Section 260

Any person authorised by the local housing authority or the Secretary of State may, at any reasonable time on giving 24 hours notice of his intention to the occupiers and to the owner if the owner is known, enter premises for the purpose of survey and examination where it appears to the authority or the Secretary of State that survey or examination is necessary in order to determine whether any powers under Part VIII should be exercised; or for the purpose of survey or valuation where the authority is authorised by that Part of the Act to purchase the premises compulsorily. An authorisation for the purposes of this section is again required to be in writing stating the particular purpose or purposes for which the entry is authorised. Section 261 creates the offence of obstructing an officer of the local housing authority, or of the Secretary of State, or a person authorised, in the performance of anything which that officer, authority or person is by this Part VIII required or authorised to do. Anyone committing this offence is liable on conviction to a fine not exceeding level 2 on the standard scale.

### 3.7.12.4 Section 319

A person authorised by the local housing authority or the Secretary of State may at any reasonable time, on giving seven days notice of his intention to the occupier and to the owner if the owner is known (see Form 9 of the Housing (Prescribed Forms) (No. 2) Regulations 1990), enter premises for the purpose of survey and examination where it appears to the authority or the Secretary of State that survey or examination is necessary in order to determine whether any powers under this Part should be exercised in respect of the premises; or for the purposes of survey and examination where a demolition or closing order, or an obstructive building order, has been made in respect of the premises; or for the purpose of survey or valuation where the authority is authorised by Part IX to purchase the premises compulsorily. As is standard an authorisation for the purposes of this section must be in writing and state the particular purpose or purposes for which the entry is authorised. It must, if so required, be produced for inspection by the occupier or anyone acting on his behalf. Section 320 creates the summary offence of intentionally obstructing an officer of the local housing authority or of the Secretary of State, or any person authorised to enter premises. Any person committing such an offence is liable on conviction to a fine not exceeding level 3 on the standard scale.

### 3.7.12.5 Section 340

A person authorised by the local housing authority may at all reasonable times, on giving 24 hours notice to the occupier and to the owner if the owner is known (see Form 9 of the Housing (Prescribed Forms) (No. 2) Regulations 1990), enter any premises for the purpose of survey and examination where it appears to the authority that survey or examination is necessary in order to determine whether any powers under this Part should be exercised.

The authorisation for the purposes of this section must be in writing stating the particular purpose for which it is given and shall if so required be produced for inspection by the occupier or anyone else acting on his behalf. By virtue of s.341 it is a summary offence to intentionally obstruct an officer of the local housing authority, or any person authorised to enter premises in pursuance of Part X, in the performance of anything which he is by that Part required or authorised to do. Anyone committing such an offence is liable on conviction to a fine not exceeding level 3 on the standard scale.

### 3.7.12.6 Section 395

Where it appears to the local housing authority that survey or examination of any premises is necessary in order to determine whether any powers under Part XI should be exercised in respect of the premises, a person authorised by the authority may at any reasonable time, on giving 24 hours notice of his intention to the occupier and to the owner if the owner is known (see Form 9 of the Housing (Prescribed Forms) (No. 2) Regulations 1990), enter the premises for the purpose of such a survey and examination.

A person authorised by the local housing authority may at any reasonable time, without any prior notice as is mentioned above, enter any premises for the purpose of ascertaining whether an offence has been committed under any of the following provisions:

- Section 346(6) (contravention of or failure to comply with provision of registration scheme)
- Section 355(2) (failure to comply with requirements of direction limiting number of occupants of house)
- Section 358(4) (contravention of overcrowding notice)
- Section 368(3) (use or permitting use of part of house with inadequate means of escape from fire in contravention of undertakings)
- Section 369(5) (contravention of or failure to comply with regulations prescribing management code),
- Section 376(1) or (2) (failure to comply with notice requiring execution of works).

An authorisation for the purposes of this section must be in writing stating the particular purpose or purposes for which the entry is authorised and must, if so required, be produced for inspection by the occupier or anyone acting on his behalf.

It is a summary offence under s.396 intentionally to obstruct an officer of the local housing authority or any person authorised to enter premises in pursuance of Part XI in the performance of anything which he is required or authorised to do. Anyone committing the offence is liable on conviction to a fine not exceeding level 3 on the standard scale.

Section 397 empowers an authorised officer to obtain a warrant for entry.

Where it is shown to the satisfaction of a justice of the peace, on sworn information in writing, that admission to premises is reasonably required either:

(a) for the purpose of survey and examination to determine whether any powers under Part XI should be exercised in respect of the premises, or

(b) for the purpose of ascertaining whether an offence has been committed under any of the provisions of Part XI (these are listed in section 395(2)),

the justice may by warrant authorise entry on the premises for those purposes or for such of those purposes as may be specified in the warrant. The power of entry conferred by a warrant includes the power to enter by force and may be exercised by the person on whom it is conferred either alone or together with others. The warrant continues in force until the purpose for which it is intended is satisfied. If the premises are unoccupied or the occupier is temporarily absent, anyone entering under the warrant must leave the premises as effectively sound against unauthorised entry as when they arrived. A justice may not grant the warrant unless satisfied

(a) that admission to the premises has been refused and, except where the purpose specified in the information is to ascertain whether an offence has been committed, that admission was sought after not less than 24 hours notice of the intended entry had been given to the occupier or
(b) that application for admission would defeat the purpose of the entry.

### 3.7.12.7 Section 600

A person authorised by the local housing authority or the Secretary of State may at any reasonable time, after giving seven days notice of his intention to the occupier, and to the owner if the owner is known, enter premises for the purpose of survey and examination where it appears to the authority or the Secretary of State that survey or examination is necessary in order to determine whether any powers under Part XVII should be exercised in respect of the premises. An authorisation must be in writing stating the particular purpose or purposes for which the entry is authorised and shall if so required be produced for inspection by the occupier or anyone acting on his behalf. Section 601(1) states that it is a summary offence intentionally to obstruct an officer of the local housing authority or of the Secretary of State, or any person authorised to enter premises in pursuance of that Part of the Act, in the performance of anything which he is required or authorised to do. Such an offence is punishable on conviction with a fine not exceeding level 3 on the standard scale.

### 3.7.13 LOCAL GOVERNMENT (MISCELLANEOUS PROVISIONS) ACT 1982

Part VIII of the Local Government (Miscellaneous Provisions) Act 1982 permits local authorities to require the registration of premises and persons connected with acupuncture, tattooing, ear piercing and electrolysis. To assist with the enforcement of this a power of entry is provided for by section 17 of the Act.

(1) An authorised officer of a local authority may enter any premises in the authority's area if he has reason to suspect that an offence under section 16 is being committed there.
(2) The power to enter may only be exercised if the authorised officer has been granted a warrant by a justice of the peace.
(3) A justice may grant a warrant only if he is satisfied that:
    (a) admission to the premises has been refused, or refusal is apprehended, or the case is one of urgency, or that an application for admission would defeat the object of entry; and
    (b) that there is reasonable ground for entry under this section.
(4) A warrant shall not be granted unless the justice is satisfied either that notice of the intention to apply for a warrant has been given to the occupier, or that the case is one of urgency, or that giving such a notice would defeat the object of entry.
(5) The warrant continues in force for
    (a) seven days, or
    (b) until the power conferred by this section has been exercised in accordance with the warrant, whichever is the shorter.
(6) When exercising the power of entry the authorised officer must produce his authority if required to do so by the occupier of the premises.
(7) Any person who without reasonable excuse refuses to permit an authorised officer of a local authority to exercise the power conferred by this section shall be guilty of an offence, and shall for every such refusal be liable on summary conviction to a fine not exceeding level 3 on the standard scale.

### 3.7.14 NOISE AND STATUTORY NUISANCE ACT 1993

Section 10 of the Act deals with audible intruder alarms and provides the means by which local authorities may adopt the provisions contained in Schedule 3 to the Act.

Paragraphs 6 to 10 of that Schedule deal with the power of entry and allow an authorised officer, on production of his authority if so required, to enter premises to turn off an alarm. The officer may not use force and may only enter if the alarm has been operating for more than one hour after it was activated and the alarm is such as to cause reasonable annoyance to those in the vicinity. Entry by force is permissible under authority of a warrant.

If a justice of the peace is satisfied on the application of an authorised officer that the alarm has been operating for more than one hour, that it is such as to give reasonable cause for annoyance to those living or working in the vicinity, that the officer has taken steps to gain access to the premises with the assistance of the keyholder and that the officer has been unable to gain access to the premises without the use of force, the justice may issue a warrant authorising forcible entry. Such a warrant will remain in force until the alarm has been turned off. Any entry exercised under this provision requires that the officer must not enter premises unless accompanied by a constable. Any officer entering unoccupied premises must, after the alarm has been turned off re-set it if that is reasonably practicable, leave a notice at the premises stating what they have done and leave the premises so far as is reasonably practicable as secured against trespassers as when he found them.

### 3.7.15 PET ANIMALS ACT 1951

Under s.4 a local authority may authorise in writing any of its officers or any veterinary surgeon or veterinary practitioner to inspect (subject to compliance with precautions specified by the authority to prevent the spread of infectious or contagious diseases among animals) any premises in their area licensed under the Act. Any such authorised person on producing his authority if so required may enter at all reasonable times such premises to inspect them and any animals on them for the purposes of ascertaining whether an offence has been or is being committed. Any person wilfully obstructing or delaying any person in the exercise of his powers commits an offence under the Act and by section 5(2) is liable to a fine not exceeding level 2 on the standard scale.

### 3.7.16 PREVENTION OF DAMAGE BY PESTS ACT 1949

Section 22 provides that any person duly authorised in writing by a local authority for the purposes of Part I of the Act, or by a person empowered by the Minister to exercise functions of a local authority under that Part may at any reasonable time enter upon any land. This entry may be for the purposes of:

(a) carrying out any inspection required by the said Part I to be carried out by the local authority;
(b) ascertaining whether there is or has been on or in connection with the land, any failure to comply with any requirement of Part I of the Act or of any notice served thereunder;
(c) taking any steps authorised by s.5 or s.6 of the Act to be taken by the local authority on or in relation to the land.

Any person duly authorised in writing by the Minister, or by a local authority to whom functions of the Minister under Part II of the Act are delegated, may, at any reasonable time, enter upon any land for the purposes of:

(a) ascertaining whether there is or has been, on or in connection with the land or any vehicle thereon, any failure to comply with any requirement of Part II of the Act or of any directions given thereunder;

(b) taking any steps authorised to be taken on or in relation to the land under Part II by a person named in an order made by the Minister or by that authority thereunder.

Anyone authorised to enter must, if required, produce evidence of his authority before undertaking any entry, and must not demand admission as of right to occupied land unless 24 hours notice of the intended entry has been given to the occupier.

Subsection 4 creates the offence of wilful obstruction which on summary conviction renders the offender liable to a fine not exceeding level 1 on the standard scale. Compensation in respect of any land damaged in the exercise of a power of entry may be recovered by any person interested in the land from the local authority on whose behalf the entry was effected.

### 3.7.17 THE PUBLIC HEALTH ACT 1936

Section 287 provides that any authorised officer of a council shall, on producing if he is so required some duly authenticated document showing his authority, have a right to enter any premises at all reasonable hours. This entry may be exercised for the purposes of:

(a) ascertaining whether there is, or has been, on or in connection with the premises any contravention of the provisions of this Act or of any bylaws made thereunder;

(b) ascertaining whether or not there are circumstances which would authorise or require the council to take any action or execute any work under the Act or byelaws;

(c) taking any action or executing any work authorised or required by this Act; or

(d) generally for the purpose of the performance of the council's functions under the Act.

Under these provisions entry to premises other than a factory or workplace may not be demanded as of right unless there has been 24 hours notice of the intended entry given to the occupier.

If it is shown to the satisfaction of a justice of the peace on sworn information in writing:

(a) that admission to any premises has been refused, or refusal is apprehended, or that the premises are unoccupied or that the occupier is temporarily absent, or that the case is one of urgency, or that an application for admission would defeat the object of entry; and that

(b) there is reasonable ground for entry into the premises,

the justice may by warrant authorise the council by any authorised officer to enter the premises, if need be by force.

A warrant should not be issued unless the justice is satisfied that the occupier has been given notice of intention to apply for a warrant or the premises are unoccupied, or the occupier is temporarily absent, or the case is one of urgency, or that such notice would defeat the object of entry.

An authorised officer entering under this section may take with him such persons as may be necessary and on leaving any unoccupied premises must ensure they are as effectively secured against trespassers as when the premises were entered.

A warrant issued under this section remains in force until the purpose for which the entry was necessary has been satisfied.

A person entering under this section must not disclose any information regarding any manufacturing process or trade secret. (Fine not exceeding level 3 on the standard scale or three months imprisonment for non-compliance.)

Section 288 creates the offence of wilful obstruction of a person in the execution of the Act, and any regulations, orders or warrants made under it. The penalty on conviction is a fine not exceeding level 1 on the standard scale.

### 3.7.18 REFUSE DISPOSAL (AMENITY) ACT 1978

Section 8 provides that any person duly authorised in writing by the Secretary of State or a local authority may at any reasonable time enter upon any land for the purpose of ascertaining whether any of the functions conferred by section 3 or 6 should or may be exercised in connection with land or for the purpose of exercising any of those functions in connection with the land.

The authorised officer must produce his authorisation if required and may not enter land as of right without first serving 24 hours notice.

Wilful obstruction is an offence and renders the offender liable to a fine not exceeding level 2 on the standard scale.

### 3.7.19 RIDING ESTABLISHMENTS ACT 1964

By s.2 a local authority may authorise in writing an officer of theirs, an officer of any other local authority, a veterinary surgeon or veterinary practitioner to inspect any premises in their area where they have reason to believe a person is keeping a riding establishment or where a licence has been granted in accordance with the provisions of this Act and is in force, or where such a licence has been applied for.

Under this Act any person authorised may, on producing his authority if so required, enter at all reasonable times any premises which he is authorised to enter. On entry he may inspect them, any horses and any thing, on or in the premises for the purposes of making a report or ascertaining whether an offence has been or is being committed. Any veterinarian authorised under this provision must be drawn from a list drawn up by the Royal College of Veterinary Surgeons and the British Veterinary Association. Any person wilfully obstructing or delaying any person in the exercise of his powers of entry or inspection commits an offence under subsection 4 and is liable on conviction to a fine of level 2 on the standard scale.

### 3.7.20 SLAUGHTER OF POULTRY ACT 1967

Section 4 of the Act provides a power of entry to premises. Any officer authorised by the local authority may enter premises. Where slaughtering is or appears to be carried on, entry may be at any time. Forceable entry is not provided for under this Act and there is no requirement for the authorisation to be in writing.

Where birds have been slaughtered within 48 hours or birds are on the premises awaiting slaughter, the powers of entry are qualified to be 'at all reasonable hours'.

Under s.4(4) it is an offence to intentionally obstruct a person in the exercise of his powers.

### 3.7.21 SLAUGHTERHOUSES ACT 1974

Under s.20 an authorised officer of a council on producing, if so required, a duly authenticated document may enter premises at all reasonable hours to ascertain possible contravention of the law and generally for the performance of the Act. Where the premises are a private dwelling entry may be demanded as of right only if 24 hours notice has been given. Under s.21 anyone who wilfully obstructs a person in the exercise of the Act commits an offence.

3.7.22 ZOO LICENSING ACT 1981

Under the Act there are three types of inspection by the Act:

- Periodical Inspections (section 10)
- Special Inspections (section 11)
- Informal Inspections (section 12).

Powers of entry are to be found in s.19 and any person who intentionally obstructs an inspector acting pursuant to this Act is guilty of an offence and liable on conviction to a fine not exceeding level 3 on the standard scale.

# Investigation in the criminal law

# 4

## 4.1 PREPARATION

An Environmental Health Officer can never know when his work will be scrutinised by others. Obviously it is subject to examination by those within his organisation, but it will on occasion also be subject to examination by others who will be less familiar with his work, less forgiving of his method and who will have, quite legitimately, high expectations of his professionalism. One such place where this expectation may be warranted is in a court of law. When an officer comes before a court or tribunal that court or tribunal will expect to be able to rely on the professionalism of the officer concerned to assist it in its deliberations. It will not be interested in excuses why a thing was not done or not done properly; it will expect the officer to have undertaken the work professionally, with attention to detail and displaying a full knowledge of all technical and legal aspects of the work in question. That is what is expected of the Environmental Health Officer and is therefore what must be, at all times, the aim. This is an ideal, and although many of us will often fail to achieve that ideal, any officer who recognises the value of such an approach will ultimately make the tasks they have to undertake so much the easier. Environmental Health Officers deal with perception. Not only must they be competent in the work they do but they must be perceived as being so by all they come into contact with. Any officer who is inattentive to detail, who fails to have regard to the appropriate method and does not keep abreast of developments will fail in this. But the officer who is aware of such needs will, when his method and credibility fall to be tested, be in a much better position to withstand that test.

Any officer who has ever given evidence in a court of law will know that it can be a lonely and frightening experience. You are not in control, this is not your normal working environment and the rules governing this activity are alien and often obscure. The only help available, should an officer get into difficulty, is that which he foresaw to prepare. He will not want to admit that he does not know the answer to a question, an answer he should know, and certainly he cannot just 'pop out' to check a fact. In short the only protection available and the only defensive weapon that can be used is that of preparation. Sound preparation is the best defence. Also because an Environmental Health Officer can never know which of the thousands of jobs they might deal with will be the one to go before a court or tribunal, the only answer is to deal with them all in the same manner. This manner must be both professional and adhere to ethical and legal standards. In a court, officers may fear cross examination, but an officer who is able to give answers modestly and carefully, who does not lose his head, who does not become aggressive and who maintains both his or her composure and account of the facts under cross examination will find the experience less fraught than expected. They will provide the opposition with little that can be used against them. Indeed it may even be that such a cross examination actually assists the officer's own case rather than that of the opposition.

How then is an officer to achieve this desirable state? There is no single answer but there are a number of means by which the Environmental Health Officer can improve his or her preparation.

## 4.2 NOTEBOOKS

It is becoming increasingly common for Environmental Health Officers to keep notebooks in the style formerly only associated with the police. These 'PACE' notebooks are something of a departure for many EHOs who have in the past used their notebook as an informal *aide-memoire* rather than a legally significant document. It is generally hard to account for the fact that the proper keeping of notebooks has not, in the past, figured prominently in the work of the Environmental Health Officer. Perhaps it is a function of the dual role of policeman and advisor, perhaps it has simply been overlooked, perhaps it was deliberately ignored. Whatever the reason the days when an EHO had the luxury of not having to keep a disciplined notebook are gone. The book, or its equivalent, is vital to the professional function of the Environmental Health Officer. Its significance in terms of the preparation of reports, statements and memory refreshing in the witness box is major. Therefore the disciplined keeping of such a book must now be considered one of the key legal skills of the Environmental Health Officer. If accepted, it is then worth considering the format of such a book. Money spent on the production and presentation of good quality notebooks is ultimately money well spent. For example, the cover. A professionally produced cover carrying the embossed coat of arms or logo of the authority, when produced by an officer, can present a much more professional image to the world generally, and the courts in particular. Such covers can be made to include a small pocket and in here can be placed the officer's identification and authorisation documents. This is important since it will ensure that any officer will, at all times and in all circumstances, carry with him these essential items. To see an officer asked to produce his authorisation document in a court, only to then see him realise he has left it in his other jacket, is an amusing thought, until it happens to you. The cover, though certainly cosmetic, will serve to improve the appearance of the book, it will look like what it is, a document having legal significance. Dirty, dogeared notebooks do not look professional and though a court should never mistake form for content it is nevertheless appropriate that form and content should be of equal quality.

The notebook itself should be large enough to be held comfortably in the palm of one hand and will be generally bound at the top of the page. The use of an elasticated page divider is useful when separating out a section of the book or finding quickly the next place where an entry is to be made. Pages must always be numbered and printed on both sides. There should be space for recording the date and time and details of inspections, incidents etc. The line spacing should be sufficient to allow the officer to use the book without having to write artificially small. This is particularly important should an officer need to refresh his memory from the book at a later date. Not to be able to read one's own writing renders the book of little value. Books should be distributed centrally and at the time they are distributed be given a unique serial number. This number should be recorded on a central register along with the name of the officer to whom it is given and the date of its distribution. It would be wise also to record the number of the book it has replaced. All old books should be verified as being full, or nearly full (number of pages remaining being recorded). Thus at any point it will be possible to ensure that all notebooks in use have been accounted for and that for any given officer it will be possible to 'track' his or her use of books from one to another, in an unbroken chain. The book itself should have space, usually at the front, to record its own number and the number of the book which preceded it and succeeds it. There should also be space to record the dates between which the book is active. Any officer can then, relatively quickly, find the relevant book for any given date or period. The name of the officer should be recorded in the book with a return address should the book be lost and later found by a member of the public. It is possible to include, in the pre-printed pages, information which may be of routine use to officers, for example the wording for the cautioning of suspects, information regarding Police and Criminal Evidence Act 1984 etc.

**Notebook Page**

| Date/Time | Details of Offence/Occurrence/Inspection/Visit etc. |
|---|---|
|  |  |
|  |  |
|  |  |
|  |  |
|  |  |
|  |  |
|  |  |
|  |  |
|  |  |
|  |  |
|  |  |
|  |  |
|  |  |

Form **4.1** Notebook page

### 4.3 MEMORY REFRESHING

As a means of recording relevant matters the notebook is vital. It is not credible to expect to recall, in adequate detail, all facts relating to an event which may have happened several months before. Also if the record of events is substantially contemporaneous then the Memory Refreshing Rule will allow any Environmental Health Officer, subsequently testifying in court, to refresh his memory in the witness box from his notebook. In a trial it is the officer's oral evidence which is of vital importance. The notebook is not evidence but is the means by which the quality, the credibility, of what the officer is saying can be enhanced. The rule on memory refreshing when giving evidence is that a witness at a trial may refresh his memory by reference to any writing made or verified by himself concerning and contemporaneously with the facts to which he testifies. It is sufficient if the writing was made or verified at a time when the facts were still fresh in the witness's memory (Attorney-General's Reference (No. 3 of 1979)). The importance of this extends beyond just the Environmental Health Officer. Any witness to any event who then records details of that event on the back of an envelope or a scrap of paper should be asked to retain this paper and to record on it the date and time that they made the observation. If there is any subsequent dispute with regard to memory refreshing from the witness statement, the witness will still have the original record to rely on. For an officer to be able to rely on a notebook in this way requires the officer to keep it correctly. The defence in any criminal case will be entitled to examine a notebook to establish that entries are consistent with the evidence being given. This can even extend to a book used for memory refreshing outside of the court. In *Owen v Edwards* (1984) a police officer refreshed his memory from his book outside of the court but did not refer to it in the court. Nevertheless it was held that where a witness for the prosecution refreshes his memory under such circumstances a defendant is entitled to examine the notebook and to cross-examine the witness on matters contained therein. Correct record keeping is therefore vital. Guidance given to the police on the completion of notebooks, and based on long experience, is entirely appropriate for use by the Environmental Health Officer.

### 4.4 DETAILS TO BE RECORDED

When using a notebook a single line should separate the current entry from the one previously made. This will clearly separate one entry from another. This is a good practice to adopt. It serves the obvious purpose of separating the current record from an earlier one and it ensures that, if subsequently referring to the record later, the officer can easily find the correct point in his notes. Each entry must be on a separate line. Any attempt to write anything previously omitted between lines or in the margins or at the bottom of a page must be avoided. This may be done perfectly innocently but may subsequently put the officer in a bad light, so it must be avoided. All entries should record the day, date and time that they were made. The timing of events should be by use of the 24 hour clock. Officers can be on duty at 9 o'clock am or pm, therefore to record an event at 21.00 hrs is much more precise. The locations of all events must be recorded. Normally this will be the full postal address of any premises. However on occasion this will not be possible; in those cases detail of location must be recorded, e.g. northern corner of field to the west of the junction of A123 and A321. The names of all persons present and all witnesses should be noted. Avoid recording 'Mr Smith ' or even 'Mr J Smith'. The full names of the person should be sought, e.g. John Michael Smith. To avoid difficulty with spellings all names should be recorded in capital letters. This will have the additional advantages of ensuring that no transcription errors occur when later drafting a statement and, if used in subsequent court proceedings, the names will be easier to find. When under pressure in a witness box the ability to rapidly and accurately find the names of parties to an incident can be of help.

If the matter concerns a company, record the name of the company secretary or clerk of the company and the address of the registered or principal office of the company. These may be required should there be a need to serve documents, e.g. notices, on the company later. Telephone and fax numbers if recorded can also save much effort later on.

The occupations and status of those present will often be needed and ought routinely to be recorded. If investigating an occurrence relating to health and safety at work for example, it is vital that the position of all within the organisation is ascertained and recorded as early as possible. Establishing and recording the status of someone as being able to speak on behalf of the company is vital. The ages of all parties can prove useful. Should any person prove to be a juvenile then the relevant provisions of the Police and Criminal Evidence Act 1984 codes will apply. A juvenile, whether suspected or not of an offence, must not be interviewed or asked to provide or sign a written statement in the absence of the appropriate adult unless Annex C of the code applies.

If there has been an offence a description of that offence should be made. Full and accurate recording involves understanding fully all elements of the offence in question; the 'points to prove' approach to offences discussed later is valuable in this regard. All documentation e.g. supplier records, safety policy, tenancy agreement should be identified and recorded for later reference. Details of any items of machinery or plant associated with events should be noted; for example this may mean recording the type of machinery concerned in any incident, its make, location, use, or the type of refrigerator, heater etc. This information should include serial numbers or other identification marks where they exist.

If a caution has been administered then this should be noted and any reply should be recorded verbatim.

It would then be appropriate to describe what the Environmental Health Officer saw, heard, did etc.

### 4.4.1 ERRORS AND VERIFICATION

Everyone at some time will make an entry in error. This is a fact and need cause little concern if dealt with correctly. It is vital at all times to remember that the notebook should be seen to have nothing to hide. Therefore no obliterations nor erasures should be made. When something has been inserted in error by an officer then such errors should be deleted by being struck through with a single line. This has the effect of clearly marking the entry as an error whilst still permitting the incorrect words to be read. From this it will be clear that the thing was entered in error and, importantly, it will be equally clear what was previously written. Pages should not be left blank. If pages are in error left blank these should be struck through and endorsed as such, i.e. write 'page left blank in error'. All such deletions should be signed or initialled and dated by the officer concerned. If this method is adopted the worst that may be said is that the officer has been a little careless but there ought not to be any grounds for a charge of deliberate bad faith on the part of the officer, such as to call into question his veracity.

All officers should be aware that the way they write their notes can undermine the credibility of any evidence they may subsequently give. An untidy notebook looks unprofessional and could be exploited by the opposition in any subsequent proceedings. A defence lawyer may use such a notebook as a means of questioning the credibility of an officer and thus of his evidence. One way round the untidy notebook problem is to write up notes after events. In Attorney General's Reference 3/79 it was held that a police officer who had taken brief jottings in the course of interviewing an accused person and within a short time thereafter made a full note in his notebook, not only incorporating those brief jottings, but also expanding them from his recollection, should be permitted to refresh his memory from that notebook at the accused person's subsequent trial.

When relying on a notebook to record events care must be taken to ensure that any record is correctly made or verified by the officer seeking to rely on the entry. In *R v Eleftheriou (Costas and Lefterakis)* (1993) the Court of Appeal found that an unverified record could not be used in evidence. In this case customs officers were concealed and working in pairs. They kept a record of observation of premises by one observing and calling out what he saw and the other writing it down. One of the officers had given evidence that, when observing, he looked up from time to time and noted his colleague recording a large number of matters and, when writing, he wrote down what was shouted to him. There had been no verification, contemporaneous or otherwise, by the observer of what the writer had written down. The court found that use by the witnesses of those records for the purposes of refreshing their memory was a material irregularity since the record had not been verified.

If challenged in open court an officer must of course make his notes available but is not required to surrender total control. For example, if requested to furnish his notebook to the defence, he may legitimately seek to control the extent of inspection by recording that, though there can be no objection to inspection of the notes being used for memory refreshing, inspection of other parts of the book ought only be done to the extent necessary to establish that the notes are kept properly. It can be pointed out to the court that the book will contain confidential information not relevant to the case in hand and it would be improper if the officer did not respect the potential sensitivity of this. Any note made otherwise than in a notebook must also be retained and this certainly extends to the use of pro-forma inspection sheets.

### 4.5 INSPECTION SHEETS

The use of such documents is widespread amongst Environmental Health departments. They are a valuable *aide-memoire* and are often used to ensure consistent standards of inspection and record keeping. Usually they present no problem with regard to contemporaneity, usually being completed as the inspection proceeds. However there may be problems if such a record fails to meet the general requirements discussed below. For example gaps left on such a form may leave an officer open to a suggestion that matters could be added later, and thus may not be a contemporaneous record.

It is essential that accuracy is observed at all times. An officer must remember it is from the observations recorded in a notebook or inspection pro-forma that any subsequent statements may be composed and decisions on enforcement taken. It is vital then that if such methods of recording information are to be used they must be completed with the same discipline as that applied to a notebook. Obtaining the highest quality of information is vital. An Environmental Health Officer should not be tempted to cut corners in any inspection undertaken; this is irrespective of the purpose considered to be behind it. Therefore an inspection report compiled with a view to enforcement procedure should be no more nor less detailed than that prepared for, for example, a house renovation grant. The aim when inspecting any premises should be to record all relevant information in such a manner that it will inform any subsequent enforcement decision, be permissible under the memory refreshing rule and allow a colleague examining the inspection findings, but who has not previously visited the premises, to be able to assess the conditions just as if he had himself visited the property. It goes without saying that all information obtained must be accurate but it must also be stressed that this information should be recorded logically and methodically to reflect a systematic and sequential form of inspection. This technique will allow for more ready retrieval of information at a later date. It is all too often the case that the apparently minor or unimportant piece of information turns out to be crucial to the quality, and ultimate successful use, of the inspection. A systematic method of inspection will reduce the likelihood of anything important being overlooked.

It is important that an officer take note of all possible sources of information. However it would be unwise to take everything that he is told at face value. For example, if told by the occupier that the property being inspected was rewired two years ago or had a D.P.C. installed last year, it would be unwise to record in the inspection findings 'the property was rewired two years ago and had a D.P.C. installed within the last year'. This without further investigation is not sustainable. If challenged not only would the officer be unable to justify these as matters of fact, they after all may result from confusion or even lies on the part of the occupier, but the officer may appear to be gullible, careless and certainly less than thorough in relying on hearsay.

As already stated a systematic approach to inspection will pay dividends both in savings of time and effort and in ensuring that a higher quality of information is recorded. Take for example the survey of a house. The noting of the information is critical – never trust to memory. All information must be recorded. Recording can be by the use of a recording device such as a cassette recorder, notebook or a pre-prepared inspection checklist. Some kind of inspection checklist or sheet can help ensure that the officer keeps to the discipline of a properly planned schedule of inspection. It also provides an immediate and obvious way of ensuring that no important element is overlooked. When recording defects either in a notebook or on an inspection sheet officers would be advised avoid the use of words such as 'Broken' or 'Defective'. Without further elaboration these do little to convey the exact nature or even extent of the problem identified. As a memory aid they are effectively useless and to try to rely on them later will inevitably prove to be futile. It should be the aim of the inspecting officer to try always to identify the element, and the nature and extent of the problem. For example, rotten woodwork to window rail, broken (or missing) sash cord to window (RHS/LHS), damp and perished wall plaster (app. 4m²) etc. It is appropriate to attempt to quantify the extent of the defect noted. To state that a room has damp-affected plaster will not without more distinguish a room which has a small damp patch, of perhaps 0.2m², from one which has an entire wall of 10m² affected. In such cases the inspection notes would be insufficient to allow all but the original inspecting officer (assuming his or her memory was good enough) to make an assessment of the property. If challenged later an officer will appear more credible if he has attempted to quantify defects. In any subsequent proceedings it is difficult to appear credible if an officer is talking about a defect but cannot speak with authority on the extent of that defect. For similar reasons as for a notebook, an Environmental Health Officer should aim to avoid leaving blank spaces on any inspection form used. If an element of the structure is sound it should be marked as such; a blank space in any inspection report might be taken to indicate that no defect was present or it might indicate that the inspector forgot to look at that particular thing. It would certainly leave the officer open to attack that, because of his record keeping, the account may not be substantially contemporaneous. It is important to note what was not inspected as well as what was. Where decorations, carpets, furniture or general clutter limit an inspection this too should be recorded. Where access was not possible to a room or other part of the premises record this, if there is a reason state what it is. This method will certainly make for a more complete inspection and add to the overall professionalism of the officer's inspection findings. Because room usage can change and to avoid confusion, rooms should be identified not only by their use but by their location e.g. kitchen (rear right-hand ground floor room). This requires the officer to be certain of, and to record, his method of orientation for identification of rooms.

If the inspection of a house is taken as an example the following can be considered the minimum information that should be recorded as part of the general information.

- The name of inspecting officer.
- The date of inspection.
- Time of day.
- Weather conditions.
- The address and location of the property.
- Approximate age of house.
- Type of construction – traditional, timber frame etc.

- The type of property: terraced, back to back, semi etc.
- The occupiers of the property. These details should include names, ages, sexes, how long they have lived there, if tenanted whose name the tenancy is in, what type of tenancy it is, what rent is paid, to whom the rent is paid, whether there is a rent book, whether the rent book was inspected.
- The type of occupancy in the house: is it an HMO, single household.
- The name and address of any managing agent and the name and address of the owner.
- Details of any mortgagee.
- Accommodation, number of rooms, type of room.
- Special features.

With appropriate amendments the same approach should be adopted for all inspections by Environmental Health Officers.

## 4.6 RECORDS KEPT BY THE PUBLIC

In addition to relying on information directly obtained an Environmental Health Officer may also have to rely on information obtained by others. It has for example become common practice for Environmental Health Officers to encourage noise complainants to keep a nuisance diary. Where this is the case it is vital that information is recorded with complete accuracy. It is therefore vital, when dealing with members of the public, that all instructions with regard to how they are to complete such documents are made clear. Firstly the potential legal significance of what the person is doing should be explained. They should be told that they may be asked to give evidence in court based on their notes. It is crucial that matters regarding memory refreshing are fully explained and clarity and precision must be emphasised from the outset. The potential witness must have explained to them both the type and significance of the information they are to record. For example they must make sure of the source of the nuisance. To have a witness stand in court and say that the noise 'must have been coming from X, where else could it have come from?', will not only possibly lose an Environmental Health Officer his case but will reflect badly on his preparation.

Complainants must record the dates, times and periods of the alleged nuisance This must be done contemporaneously with events. It is a good idea to ensure that both start times and finishing times are where possible recorded. If the person is going to be away for any reason ensure that they record this. This will explain any 'quiet' period revealed by the form. If any photographs are taken to support entries the complainant must be told to retain all negatives. They must record the date and time of all photographs taken and they must date, time and sign the reverse of all prints accordingly. They should be encouraged to record as much information as possible on the type of nuisance alleged. If it is dust what colour was it, did it have a taste? What was the effect? If it was a noise what type of noise: high pitch, low pitch, impulsive, music, shouting etc. They should be encouraged to record how the nuisance affected them. Where appropriate they should be encouraged to use sketches to clarify location, orientation etc. They may wish to use a scale to help quantify their response, e.g. 5 = Very annoying, 1 = Barely discernible, or be more descriptive, e.g. Barely audible, loud, very loud, deafening. To assist in recording such information an example of a Record of Nuisance is reproduced. If such a document is used it would be good practice to both explain and complete a specimen entry in order that any person recording information has an example of what is expected of them. Any officer proposing to use such a sheet must, however, make it very clear to any potential complainant that the purpose of this document is not to canvass for complaints, but simply to ensure that the details of any complaint, that would have ordinarily been made, are recorded in an appropriate and accurate way. Any Environmental Health Officer involved in an investigation which, in part, relies on records kept by the public must understand that the keeping and presentation of these records alone is not enough. They do not free the officer of the attendant obligation to undertake all enquiries necessary to satisfy himself of the justification of any complaint.

### 4.7 PORTABLE TAPE RECORDERS

It is known that some Environmental Health Officers are in the habit of carrying portable tape recorders when undertaking their duties. There remains however some uncertainty as to the propriety of doing so. In 1991 Wolchover and Heaton-Armstrong[1] recorded that, for police officers:

> We can see no legal objection to the use of hand held tape recorders and suggest that prudent officers might start using them now without waiting for approval from above. For those who, understandably, do not feel sufficiently courageous to follow this course, we advocate a much more cautious approach to the endorsement by suspects of written interview records compiled after the event.

The Royal Commission on Criminal Justice Research Study 22 on the Questioning and Interviewing of Suspects Outside the Police Station, found that approximately 6% of police officers, and possibly more, were already carrying personal tape recorders. The study felt that, for the police, the use of portable tape recorders would represent a positive step towards improving the treatment of suspects and should therefore be

# NUISANCE RECORD

Source of nuisance (address/location/name): ............................................................................

Type of nuisance: ..................................................... Record kept from: ........................to ....................(dates)

Record kept by: (names) ...........................................(address) ....................................................................

From (start date)............................................................ to (end of record date)................................................

Number of pages.........................

I/We certify that the following entries are a true and contemporaneous record of events

Signed...................................................................... Dated.......................

**The first line of this form has been completed as an example for you to follow when recording your own information**

| Date & Day and | Time Start | Time End | Source of disturbance | Weather conditions | Effect of disturbance (e.g. sleep interference, headaches, prevention of enjoyment of garden, damage) | Location of person recording alleged nuisance | Signed Inc. Date and Time |
|---|---|---|---|---|---|---|---|
| Mon. 1/1/99 | 21.00 | 23.45 | 1 Train Lane, Music, heavy beat | Fine and dry | Could not hear television at normal volume | Sitting Room, 3 Train Lane, windows closed | J Smith 1/1/99, 23.46 |
| | | | | | | | |
| | | | | | | | |
| | | | | | | | |
| | | | | | | | |
| | | | | | | | |
| | | | | | | | |

Form 4.2 Nuisance record

encouraged. This was because it was felt that this could improve the quality and validity of interviews conducted inside the police station. A central aspect of the investigative process, i.e. the general process of questioning and discussion leading to an officer's having reasonable ground to believe an offence had been committed, would then be less obscure. The study did not suggest that recordings made under such circumstances were meant to replace formal interviews inside the station. But the study did record that interviewing is not a one-off event, but a process, and the technology is available for full documentation of that process to be provided. The study recommended that the use of portable tape recorders should not be confined to interviews. Early questioning and other non-offence related conversations with a suspect should also be recorded and the tape recording of conversations with witnesses was recognised as of value. There seems to be much in this study that could apply to the work of the Environmental Health Officer and properly used portable recording devices could have an important role to play in the work of the Environmental Health Officer.

### 4.8 THE CRIMINAL PROCEDURE AND INVESTIGATIONS ACT 1996

The Act introduces new provisions with regard to criminal procedure and criminal investigations and impacts particularly on the need to record information and on the use to which such information may be put. The Act is divided into seven parts. For the Environmental Health profession Parts I and II are the most significant. Part I of the Act deals with the requirement for the disclosure of unused prosecution material and Part II relates to a code of practice for the conduct of criminal investigations. Section 1 of the Act makes it clear that the provisions relating to disclosure apply to criminal investigations. For the purposes of the Act a criminal investigation is an investigation which police officers or other persons have a duty to conduct with a view to its being ascertained:

(a) whether a person should be charged with an offence, or
(b) whether a person charged with an offence is guilty of it.

This reference to 'other persons' would appear to apply to Environmental Health professionals in the same way as the Police and Criminal Evidence Act 1984 provisions. Generally the provisions of Part I apply to both summary trials and trial on indictment. The obligation to disclose material to the defence is not however one which will fall on the Environmental Health professional and is best left in the hands of the relevant solicitors. In Part II of the Act, at s.23, provision is made for the making of a code of practice for criminal investigations. Part II of the Act would, on the face of it, appear not to be relevant to the work of an Environmental Health Officer as it is limited, by s.22(1), to criminal investigations by police officers only. But s.26(1), when considering the effect of the code, states that a person other than a police officer who is charged with the duty of conducting an investigation with a view to its being ascertained:

(a) whether a person should be charged with an offence, or
(b) whether a person charged with an offence is guilty of it,

shall in discharging that duty have regard to any relevant provision of a code which would apply if the investigation were conducted by police officers. Clearly then, though an Environmental Health professional may not have to have regard to those provisions which could apply only to the police, he or she must have regard to all other parts of the code. The code deals with the general responsibilities of an investigator, the recording of information, the retention of material, the preparation of material for a prosecutor and the disclosure of material to an accused. It is instructive to note here that the provisions of the code require that if information is obtained which may be relevant, it must be recorded at the time it is obtained or as soon as practicable after that time and that in some cases negative information may be relevant to the investigation and should be recorded. An example of negative information would be when a series of people present in a particular place at a particular time state they saw nothing unusual. The code requires an investigator to retain material obtained in an investigation. The material to be retained includes not only material which may come into the possession of an investigator (papers etc.) but also material generated by him (interview records etc.). The code identifies that the duty to retain material includes in particular the duty to retain material falling into the following categories:

- crime reports (including crime report forms, relevant parts of incident report books or police officers' notebooks);
- records of telephone messages containing descriptions of an alleged offence or offender;
- final versions of witness statements (and draft versions where their content differs in any way from the final version), including any exhibits mentioned;
- interview records;
- communication between police and experts, reports of work carried out by experts and schedules of scientific material prepared by the expert;
- any material casting doubt on the reliability of a confession;
- any material casting doubt on the reliability of a witness.

## 4.9 POINTS TO PROVE

When recording information, and later when taking statements, it is vital that the officer has in mind at all times the elements of any potential offence he may be investigating. This is the so called 'points to prove' approach to offence investigation. This approach is well recognised by police officers[2] but has not traditionally been an approach overtly used by Environmental Health professionals despite the applicability to many areas of their work. The prosecution in any criminal case must generally prove two things: firstly *actus reus*, i.e. that all the elements constituting the offence charged were brought about by the conduct of the defendant, and secondly *mens rea*, i.e. that the defendant brought about those elements with a particular state of mind. The burden in English law is on the prosecution to prove all of these. Whatever the crime the prosecution must prove every part of the *actus reus*. There is no concept of 'near enough'. Near enough is not good enough and the burden is on the accuser to prove all elements of his accusation (*Woolmington v DPP* [1935]). Generally in English criminal law whoever bears the burden of proof on an issue will lose on that issue if after presenting their case the jury or magistrates retain sufficient doubt as to whether that issue has been established. Thus each and every element of an offence must be established to the criminal standard, 'beyond a reasonable doubt', if the case is to be proven against an accused. It is therefore vital that any investigating officer recognise precisely what it is that he is required to prove and use this knowledge to direct his investigations and seek to obtain evidence of those matters. This is perhaps easier to explain by way of examples. Consider the Food Safety Act 1990, s.14 and the Housing Act 1985, s.369.

*Section 369 Housing Act 1985*

This section provides that it is an offence for a person managing an HMO to knowingly contravene or without reasonable excuse fail to comply with a regulation under s.369. Against an accused the 'points to prove' can be broken down as follows.

**That you.** The prosecution must be able to prove that the person before the court is the person they are alleging has committed the offence.

**Being the person managing.** This requires reference to s.398 of the Housing Act 1985 and r.2(1) of the Housing (Management of Houses in Multiple Occupation) Regulations 1990 for the definition of person managing. Care should be taken to recognise the changes to the definition of person managing introduced by the Housing Act 1996.

**The House in Multiple Occupation.** A house in multiple occupation is a house occupied by persons who do not form a single household (s.345, Housing Act 1985) (as extended by the addition of subsection 2, Local Government and Housing Act 1989). The need to establish all elements of this definition is essential.

**House.** Section 399 of the Housing Act 1985 gives a definition for the purposes of the Act that 'house' includes any yard, garden, outhouses and appurtenances belonging to the house or usually enjoyed with it. Regard should also be had to decided cases e.g. *Critchell v Lambeth B.C.* [1957], *Quillotex v Ministry of Housing and Local Govt.* [1966], *Okereke v Brent LBC* [1966].

**Occupied.** This has been held to a question of fact in each case. Once again decided cases may need to be considered. *Minford Properties v LB of Hammersmith* (1978), *Silbers v Southwork LBC* (1977).

**Form a Single household.** This is often the key question. Though guidance is available from government circulars and professional guidance notes there remain no sure criteria. Thus it is here more than anywhere that regard must be had to decided cases, eg. *Hackney LBC v Ezedinma* [1981] *Simmons v Pizzey* [1979].

**Did knowingly contravene.** If this is what is alleged it will require proof of *mens rea* i.e. guilty knowledge; the offence is not simply contravening but knowingly contravening. The alternative may be alleged.

**Did without reasonable excuse fail to comply.** Normally it may be expected that the accused would have to establish reasonable excuse (s.101 of the Magistrates Court Act 1980), but see *City of Westminster v Mavroghenis* [1983].

**With a regulation under s.369, Housing Act 1985.** The precise regulation(s) allegedly contravened or not complied with must be identified, and evidence of this must be presented.

*Section 14 Food Safety Act 1990*

Any person who sells to the purchaser's prejudice any food which is not of the nature or substance or of the quality demanded by the purchaser, shall be guilty of an offence.

The first point an officer should note is that this is an offence of strict liability (*Betts v Armstead* (1888)) and there is thus no need to prove *mens rea*. The officer need not invest time in seeking to establish the mental element of this offence as none is needed. Officers may wish to present evidence of culpable neglect as an aggravating feature of the offence but this is relevant to sentence not to establishing guilt or innocence. An officer is required only to establish the *actus reus* of the offence. Proof of the *actus reus* of this offence requires the officer to establish:

(a) that a person
(b) has sold
(c) to the purchaser's prejudice
(d) food
(e) which was not of the nature or
(f) which was not of the substance or
(g) which was not of the quality
(h) demanded by the purchaser.

If the 'points to prove' approach is considered in more detail then each element falls to be considered in turn as follows.

**That a person.** The person who made the sale would have to be identified. Again the prosecution must be able to prove that the person before the court is the person they are alleging has committed the offence. This would require the officer to obtain statements from the purchaser identifying the person charged with this offence as the person who actually sold the food. Till receipts etc if available would make useful exhibits. This would apply equally to the situation where the action was to be taken against a large national retailer.

**Has sold.** Selling would normally involve the transfer of property and requires an officer to establish that such a transfer has taken place. Statements again would need to be taken to establish the occurrence of a sale. By s.2 of the Food Safety Act 1990 the term sale is given an extended meaning under the Act to include any supply otherwise than on sale, in the course of business. The investigating officer must therefore establish the sale as being within the normal definition of sale or within the extended definition contained within s.2. This may mean obtaining statements not simply from the 'purchaser' of the food but also from the 'vendor' as to the circumstances under which the food was supplied. It is important to remember that in s.14(2) the reference to sale is to be construed as a reference to sale for human consumption. There would need to be evidence that such was the case. Any investigating officer should remember that s.3(2) provides that any food commonly used for human consumption shall, if sold or offered, exposed or kept for sale, be presumed, until the contrary is proved, to have been sold or, as the case may be, to have been or to be intended for sale for human consumption. There would be a requirement, if relying on this subsection, to establish that the food is commonly used for human consumption. Again statements could be taken to establish this. The officer should know that in this section 'Sell food' means to sell as food (*Meah v Roberts* [1978]); thus were the person to purchase something which was not food but which had been sold as food it would come within this section providing there was evidence of its having been so sold.

**To the purchaser's prejudice.** The officer would then need to establish that the sale had been to the 'The Prejudice of the Purchaser'. Clearly the purchaser would have to be identified as such. Again statements will inevitably cover this. Prejudice would then need to be established. A purchaser cannot be prejudiced when clear and unambiguous notice is given to and received by him at the time of the sale that the article is not of the nature, substance or quality he demands. The purpose of this area of law is not to protect a person from making a bad bargain. Any investigating officer must therefore seek to establish the circumstances surrounding the sale and specifically if any notice was given. The officer will thus have to enquire about such matters as any notice given to the purchaser regarding the food. If there was notice enquiries will have to be directed as to how that notice was given, and whether it was clear and unequivocal (*Collet v Walker* (1895)). How was the purchaser prejudiced? Prejudice in this context requires only that the food was inferior to that requested and paid for. The officer's investigations must reveal with clarity and precision what was requested and paid for.

**Food.** Section 1 of the Food Safety Act 1990 defines food to include: drink, articles or substances of no nutritional value which are used for human consumption, chewing gum and other products of like nature and use, and articles and substances used as ingredients in the preparation of food. Though the definition is unlikely to pose a serious problem the officer should remember that his investigations should establish that the thing sold was in fact food.

**Which was not of the Nature, Substance and Quality demanded by the purchaser.** There is an area of common ground between the three words: nature, substance and quality (*Preston v Greenclose Ltd.* (1975)). It should however usually be possible to decide which of the three words is the most appropriate. It is up to the investigating officer to ascertain which it is he is dealing with and to accumulate evidence accordingly. In any proceedings a summons should not be for selling food 'not of the nature, substance or quality', Since this wording alleges three offences and is bad for uncertainty.

Even after establishing all the 'points to prove' an investigating officer must recognise that there are statutory defences which may be put forward and his investigations should encompass these, so far as is practicable. In adopting this approach it becomes clear how important it is that any investigating officer, once he is considering enforcement action, ensures that he fully understands all elements of the offence in question and uses that knowledge to guide him in the undertaking of any enforcement action. Any omission to establish a key point will prove fatal to the officer's case and of course be personally embarrassing.

### NOTES

1 Wolchover and Heaton-Armstrong (1991) Cracking the Codes, *Police Review*, 12 April, 751.
2 Calligan S, 1995. *Points to Prove*, 4th Edition, Police Review Publishing Co. Ltd., London.

# Police and Criminal Evidence Act 1984

# 5

## 5.1 HISTORY

If there was one statutory provision which perhaps passed initially almost unnoticed by Environmental Health Officers but was to become central to much of their enforcement activity it was the Police and Criminal Evidence Act 1984 (PACE). This Act introduced legislation having as its aim the regulation of the way in which persons, suspected of a criminal offence, were detained and questioned by the police and others. The Act was a response to a number of concerns regarding this area of activity and was intended to provide protection both for those subject to investigation and for those carrying out that investigation. It would of course be wrong to think that this area had gone entirely unregulated before the 1984 Act. The provisions of the Police and Criminal Evidence Act 1984 replaced the earlier guidance found in the 'Judges' Rules', which had regulated practice in this area. These had, since 1912, regulated, though not on a statutory footing, the detention, arrest, search and questioning of suspects by the police and others. The Rules stated that they were never intended to affect the overriding principles of law in this area but they did emphasise that a breach might render information so gained inadmissible in subsequent proceedings. The 1984 Act changed this. Observation of the provisions of the Police and Criminal Evidence Act 1984 will now operate so as to ensure the admissibility of evidence acquired, fairness to all suspects and that all Environmental Health Officers follow the same procedures wherever they are in the country.

## 5.2 APPLICATION

Because of the title it might be thought that that this statute relates only to criminal matters investigated by the police. This would be wrong. Section 67(9) of the Act stipulates that persons other than police officers who are charged with the duty of investigating offences or charging offenders shall in the discharge of that duty have regard to any relevant provision of any code issued under s.66. The provisions of the Act have therefore been held to apply to customs officers (*R v Weerdesteyn* (1994)), store detectives (*R v Bayliss* (1993)) and RSPCA inspectors (*Stilgoe v Eager* (1994)). There is no doubt that they apply to an Environmental Health Officer.

Section 67(11) of the Act goes on to provide that in all criminal and civil proceedings any code issued under s.66 of the Act shall be admissible in evidence; and if any provision of such a code appears to the court or tribunal conducting the proceedings to be relevant to any question arising in the proceedings it shall be taken into account in determining that question. Section 67(10) also makes it clear that a failure on the part of any person, other than a police officer, who is charged with the duty of investigating offences or charging offenders to have regard to any relevant provision of such a code in the discharge of that duty, shall not of itself render him liable to any criminal or civil proceedings.

### 5.3 CODES OF PRACTICE

Section 66 of the Act requires that the Secretary of State shall issue codes of practice in connection with –

(a) the exercise by police officers of statutory powers –
  (i) to search a person without first arresting him; or
  (ii) to search a vehicle without making an arrest;
(b) the detention, treatment, questioning and identification of persons by police officers;
(c) searches of premises by police officers or on persons premises.

The most recent codes issued under section 66 of the Act, taking account of changes introduced by the Criminal Justice and Public Order Act 1994, came into effect in April 1995 and regulate activity in the following areas:
A. The exercise by police officers of statutory powers of stop and search
B. The searching of premises by police officers and the seizure of property found by police officers on persons or premises
C. The detention, treatment and questioning of persons by police officers
D. The identification of persons by police officers
E. The tape recording of interviews by police officers at police stations with suspected persons.

Of particular importance to the Environmental Health Officer are Codes of Practice B, C, and E.

Under the Act (s.78) a court may in any proceedings refuse to allow evidence on which the prosecution proposes to rely to be given if it appears to the court that, having regard to all the circumstances, including the circumstances in which the evidence was obtained, the admission of the evidence would have such an adverse effect on the fairness of the proceedings that the court ought not to admit it. A failure to observe the provisions of a code could therefore result in evidence being excluded. Take for example the interviewing of a youth in breach of the codes of practice. To do so will mean that any evidence of any confession obtained during the interview will be excluded under section 76 of the Police and Criminal Evidence Act 1984 (*R v W and Another* [1994]). There is no inevitability about the exclusion of evidence obtained in disregard of the codes. It is in all cases for the court to decide if it will allow evidence obtained otherwise in accordance with the codes. It would, however, be foolhardy to deliberately seek to ignore the provisions of any relevant code.

## 5.4 CODE B

Code B of the Police and Criminal Evidence Act 1984 codes applies to searches of premises:

1.  Undertaken for the purpose of an investigation into an alleged offence, with the occupier's consent, other than searches made which are:
    (a) routine scenes of crime searches
    (b) calls to a fire or burglary made by or on behalf of an occupier or searches following the activation of a burglar alarm
    (c) bomb threat calls
    (d) where it is reasonable to assume that that an innocent occupier would agree to and expect that police should take the proposed action.
2.  Under powers conferred by ss. 17, 18 and 32 of the Police and Criminal Evidence Act 1984
3.  Undertaken in pursuance of a search warrant.

Section 17 of the Police and Criminal Evidence Act 1984 allows a constable to enter and search premises for the purposes of:

(a) executing a warrant of arrest or commitment,
(b) arresting a person for an arrestable offence and offences under certain statutes,
(c) recapturing an escapee, or
(d) saving life or limb or preventing serious damage.

Section 18 allows a constable to enter and search any premises occupied or controlled by a person who is under arrest for an arrestable offence if he has reasonable grounds for suspecting that there is on the premises evidence relating to that offence. Section 32 allows a constable to enter and search any premises in which an arrested person was when arrested or immediately before he was arrested, for evidence relating to the offence for which he has been arrested.

An Environmental Health Officer has some of the most extensive investigative powers available to any public official. Though, as already examined, there are variations, typical of the powers granted are those found in s.108 of the Environment Act 1995. These allow an Environmental Health Officer at any reasonable time (or, in a situation which his opinion is an emergency, at any time) to enter premises which he has reason to believe it is necessary for him to enter. On entering he may make such examination and investigation as may in any circumstances be necessary; including the taking of such measurements and photographs and recordings as he considers necessary for the purpose of any examination or investigation. In the case of any article or substance found in or on any premises being an article or substance which appears to him to have caused or to be likely to cause pollution of the environment, he may take possession of it and detain it for so long as is necessary for a number of purposes. All of the above powers are exercisable independent of the existence of any suspicion by an officer of any offence having been committed. They may be available on a routine visit. However the case of *Dudley MBC v Debenhams* (1994) found that a routine inspection by a Trading Standards Officer (equally applicable to an Environmental Health Officer) could nevertheless be a search within the provisions of the Police and Criminal Evidence Act 1984 codes. In the *Dudley* case Trading Standards Officers entered a shop, as of right, under powers conferred on them by section 29 of the Consumer Protection Act 1987. Under that Act, though they had power to enter, they did not have powers subsequently to require an employee to provide business records, including a computer print record. The officers were there only to ascertain whether there was evidence of any contravention of the 1987 Act. Thus it was that when they asked for a computer printout they were dependent on the employee's consent being given willingly, they had no powers to compel this. The court felt that an employee in that situation ought to have the benefit of the protection of Code B. The code required that there be given a Notice of Powers and Rights of search and required that in such circumstances the employee be told that he was not obliged to consent to the request. The court found that, for the purposes of PACE, a search took place when a person entered and looked about. It was not necessary that there should be any physical interference with goods. The officers were under a duty to enforce compliance with 1987 Act, having in mind an intention to prosecute if appropriate. Such an activity was a search and therefore Code B should apply. The court felt it was clear that the provisions of the code could very well be complied with at the moment an officer entered premises, albeit unexpectedly. It was envisaged that an officer would have to have with him a notice of his powers, the rights of the occupier and would have to issue the relevant warning concerning consent. These he could do as soon as he arrived on the premises. To achieve this it was accepted that an officer would have to carry a suitably drafted notice to present to an occupier This was considered acceptable as none of the provisions of Code B required that advance warning had to be given. This, though a significant finding, is not so extensive a ruling as might at first appear. The newest version of Code B makes it clear, at para. 1.3B, that the code does not apply to the exercise of a statutory power to enter premises, or to inspect goods, equipment or procedures, if the exercise of that power is not dependent on the existence of grounds for suspecting that an offence may have been committed and the person exercising the power has no reasonable grounds for such suspicion. Therefore where it is intended to search premises under the authority of a warrant or a power of entry which is not dependent on suspicion of an offence (i.e. a routine visit) Code B need not apply. Where the code does apply it stipulates that if the search is to be one with the consent of the person entitled to grant entry to the premises, that consent must, if practicable, be given in writing on a Notice of Powers and Rights before the search takes place. Any officer must make enquiries to satisfy himself that any person giving consent is in a position to give such consent. In the case of, for example, an HMO a search should not be made having only obtained the landlord's consent unless the tenant is unavailable and the matter is urgent. Before seeking consent the officer must state the purpose of the proposed search and inform the person concerned that there is no obligation to consent. He must also inform them that anything seized may be produced in evidence. If at the time of the proposed search the person is not suspected of an offence the Environmental Health Officer must tell him this when stating the purpose of the search. It is not necessary to seek consent where in the circumstances it would cause disproportionate inconvenience to the person concerned.

### 5.4.1 NOTICE OF POWERS AND RIGHTS

Paragraph 5.7 of the code requires that an officer who conducts a search must, unless it is impracticable, give copy of Notice of Powers and Rights before the search takes place. Such a notice must contain information:

- Specifying whether the search is made under warrant, or with consent, or in the exercise of powers under ss. 17, 18 and 32 of the Police and Criminal Evidence Act 1984. The format of any notice is required to provide for authority or consent to be indicated where it is appropriate.
- Summarising the extent of the powers of search and seizure contained in the Act.
- Explaining the rights of the occupier and the owner of the property seized set out in the Act and in the code.
- Explaining that compensation may be payable in appropriate cases for damage caused in entering and searching.
- Stating where a copy of the stop and search code may be consulted.

A specimen of such a Notice of Powers and Rights is reproduced in LACOTS Circular FS 7 95 5, May 1995 and officers would be advised to note its content. The provisions regarding ss.17, 18 and 32 of the Police and Criminal Evidence Act 1984 do not apply to an Environmental Health Officer and therefore any notice used would have to be appropriately altered to conform with the requirements of the code. A typical first page of such a notice is reproduced here. It must be recognised however that this would be only the first page of the notice and should not be used alone without the accompanying summary of rights.

Environmental Health Officers will, it appears, therefore find themselves engaged in two kinds of enquiry. The first will be where the officer is engaged on a routine visit. This will not be a search for the purposes of Code B. Where an officer is using powers, such as those in Environment Act 1995 or Health and Safety at Work etc. Act 1974, s.25., which are not dependent on that officer having reasonable grounds for suspecting an offence has been committed, then by virtue of para. 1.3B of Code B the provisions of the code do not apply. However a visit where an Environmental Health Officer has a power of entry but also has reasonable ground for suspecting that an offence has been committed (for example where information or complaint has been received) would appear to be a search regulated by Code B and its provisions will therefore have to be followed. In such cases it would be either a search with consent or a search without consent. Provisions must be followed accordingly. In this sense it would be similar to a search by a police officer under ss.17, 18 and 32 of the Police and Criminal Evidence Act 1984. It is important to note that the phrase used is a 'search with consent' and not a search 'with the need for consent'. This must be taken to refer to an existing fact and not to any condition precedent. The codes must envisage a set of circumstances where an officer may have a power of entry and search irrespective of consent. Clearly a search can be undertaken with or without consent depending on circumstances. Code B, at para.

Premises Search Record Ref: ..............................

## NOTICE OF THE POWERS TO SEARCH PREMISES AND OF THE RIGHTS OF OCCUPIERS.
### Police and Criminal Evidence Act 1984

Address of Premises: ...............................................................................................
.......................................................................................................................................

**POWER UNDER WHICH SEARCH MADE**
☐ **warrant/order** (copy attached)

or ☐ **Statutory Power**.
..................... (Name of officer) is authorised under the following provisions to enter premises and to examine items thereon as provided for under, and in accordance with, those enactments in the exercise and discharge of the functions of the Council. This search is being undertaken under the provisions indicated.

### Schedule of Statutory Provisions

☐ Animal Boarding Establishments Act 1963
☐ Animal Health and Welfare Act 1984
☐ Breeding of Dogs Act 1973, 1991
☐ Building Act 1984
☐ Building Regulations 1985
☐ Caravan Sites and Control of Development Act 1960
☐ Clean Air Act 1993
☐ Control of Pollution Act 1974
☐ Dangerous Wild Animals Act 1976
☐ Environment Act 1995
☐ Environmental Protection Act 1990
☐ Food Act 1984
☐ Food Safety Act 1990
☐ Health and Safety at Work etc. Act 1974
☐ Housing Act 1985

☐ Local Government and Housing Act 1989
☐ Local Government (Miscellaneous Provisions) Act 1976 & 1982
☐ Offices, Shops and Railway Premises Act 1963
☐ Pet Animals Act 1951
☐ Prevention of Damage by Pests Act 1949
☐ Public Health (Control of Disease) Act 1984
☐ Public Health Act 1936, 1961
☐ Refuse Disposal (Amenity) Act 1978
☐ Riding Establishments Act 1964, 1970
☐ Shops Act 1950
☐ Slaughter of Poultry Act 1967
☐ Slaughterhouses Act 1974
☐ Zoo Licensing Act 1981

or ☐ **With written consent** of person entitled to grant entry (only applicable where above powers are not exercised).
I hereby consent to ................ (Name of Officer/s) searching the above named premises.
Signature: ...............................................................................................................
Name (Capitals): ........................................................... Age: ...................................
Status relative to premises: ........................................................................................

**Officer in Charge of Search**
Name: ................................................................ Designation: ..................................
Authority and Department: .........................................................................................
Address: ...................................................................................................................
.................................................................................... Postcode: ....................

| Date/Time of search: |
| --- |

**Form 5.1** Notice of the powers to search premises and of the rights of occupiers

5.4, makes it clear that a search with consent must be preferred, as it states that any officer intending to enter otherwise than with consent should first attempt to communicate with the occupier and ask that person to allow him to enter, i.e. make it a search with consent. So it appears that any EHO who conducts a search with or without consent, where there is suspicion that there may have been an offence, must give a Notice of Powers and Rights, and that notice must identify if the search was made under warrant, under statutory powers or with consent. If at any point an officer engaged in a routine visit acquires reasonable grounds for suspecting an offence has been committed then he may be obliged to treat the situation as being one of a search regulated by Code B and issue the appropriate Notice of Powers and Rights. Where an officer finds any consent to search premises is withdrawn then he must not continue to search those premises. Such a withdrawal of consent may, as has been seen, constitute an offence of obstruction under the relevant statutory provision and action should be considered accordingly. If the entry was obtained under warrant and permission for entry had been obtained but subsequently revoked, this would not have any practical effect as this entry and search was undertaken by authority of the warrant and not by permission. It is perhaps important for the officer to emphasise that, even though for courtesy's sake permission may be sought, when entering he is doing so under authority of the warrant not of the permission.

### 5.4.2 RECORD OF SEARCH

Where premises have been searched Code B requires the officer who was in charge of the search to make a record of that search. Such a record must include:

1. The address of the premises searched.
2. The date, time and duration of the search.
3. The authority under which the search was made. Where the search was made in the exercise of a statutory power to search premises without warrant, such a record must include the power under which the search was made; and where the search was made under warrant or with written consent a copy of the warrant or written consent must be attached to the record or kept in some place which is identified in the record.
4. The names of all officers who conducted the search.
5. The names, if known, of any people on the premises.
6. Either a list of any articles seized or a note of where such a list is kept. If the seizure was not covered by a warrant the reason for any seizure.

7. Details of any damage caused during the search and the circumstances surrounding that damage.

Where premises have been searched under warrant, that warrant must be endorsed to show:

1. Whether any articles specified in the warrant were seized.
2. Whether any other articles were seized.
3. The date and time at which it was executed.
4. The names of the officer(s) who executed it.
5. Whether a copy of the Notice of Powers and Rights was handed to the occupier or whether it was endorsed as required by paragraph 5.8 of Code B and left on the premises. If it was the record should note where it was left.

LACOTS is on record as believing that the provisions of Police and Criminal Evidence Act 1984 Code B will apply in only relatively limited circumstances (LACOTS Circular FS 7 95 5, May 1995) i.e. where a search is being made with regard to an alleged offence and an officer is acting outside his/her statutory authority. LACOTS does not consider the use of Code B notices necessary except for cases when an officer cannot point clearly to a power authorising him/her to act without consent. This is a valid alternative view but does run contrary to existing precedent and the strict wording of the codes. Individual Environmental Health Officers must decide for themselves which course of action to follow.

## 5.5 CODE C

Code C deals with the detention, treatment and questioning of persons by police officers and is, inter alia, designed to protect against unfairly obtained information. Particularly it is paragraph 10 and subsequent paragraphs which are of interest to the EHO.

### 5.5.1 CAUTIONS

Paragraph 10 of Code C stipulates that a person, whom there are grounds to suspect of an offence, must be cautioned before any questions about it (or further questions if it is his answers to previous questions that provide grounds for suspicion) are put to him regarding his involvement or suspected involvement in that offence, if his answer or his silence may be given in evidence to a court in a prosecution. It is therefore not necessary to caution a person when speaking to him for other purposes; for example, to establish his identity or his ownership of any article or seek verification of a written record. Also whenever a person, who is not under arrest, is initially cautioned or is reminded that he is under caution, he must at the same time be told that he is not under arrest and is not obliged to remain with the officer. Under rule II of the old Judges' Rules 1964 a caution had to be given: 'As soon as an officer had evidence which would afford reasonable grounds for suspecting that a person has committed an offence before putting to him any questions relating to that offence.' In *R v Osbourne & Virtue* [1973] it was held that an officer was not bound to caution until the point when he had got some information which he could put before a court as the beginnings of a case. As such it appeared an objective test. However an officer is now obliged to caution when he has grounds (not reasonable grounds) for suspecting a person has committed an offence. As an apparently subjective test, those grounds do not have to amount to evidence, but if an officer merely suspects that the person has committed an offence he is entitled to question without first cautioning (*R v Shah* (1994)). It seems clear therefore that the aim here is to ensure that a caution is administered early. This view is further reinforced by the changes introduced by the Criminal Justice and Public Order Act 1994. The caution prior to the introduction of the Criminal Justice and Public Order Act 1994 was in the following terms: 'You do not have to say anything unless you wish to do so, but what you say may be given in evidence.'

Minor deviations from this formula did not constitute a breach of the requirement provided that the sense of the caution was preserved. The provisions of s.34 of the Criminal Justice and Public Order Act 1994 have had the effect of requiring the introduction of a new form of caution and because the provisions of s.34(4) of the 1994 Act extend the provisions of s.34(1) to situations when a person is questioned by anyone charged with the duty of investigating offences, it is clear that all persons charged with such a duty, including Environmental Health Officers, must use the new form of caution. Section 34 of the Act now permits 'proper inferences' to be drawn from a suspect's silence. The section states that where in any proceedings against a person for an offence evidence is given that the accused:

- at any time before he was charged, on being questioned under caution by a person trying to discover whether or by whom the offence had been committed, failed to mention any fact relied on in his defence in those proceedings, or
- on being charged with the offence or officially informed that he might be prosecuted for it failed to mention any such fact, being a fact which in the circumstances existing at the time the accused could reasonably have been expected to mention when so questioned, charged or informed as the case may be, the court, in determining whether an accused is guilty of the offence charged, may draw such inferences from the failure as appear proper.

For this reason it has been necessary to rephrase the caution so as to make it clear to the accused what the significance of their silence may be.

### 5.5.2 THE NEWEST CAUTION

The newest caution takes the following format: 'You do not have to say anything. But it may harm your defence if you do not mention when questioned something which you later rely on in court. Anything you do say may be given in evidence.'

As before, minor deviations from this formula do not constitute a breach of the requirement provided that the sense of the caution is preserved. It is important to note that this does not amount to an abolition of the right to silence, a person is still not compelled to speak. Indeed there are a number of limitations on s.34 that make this quite clear.

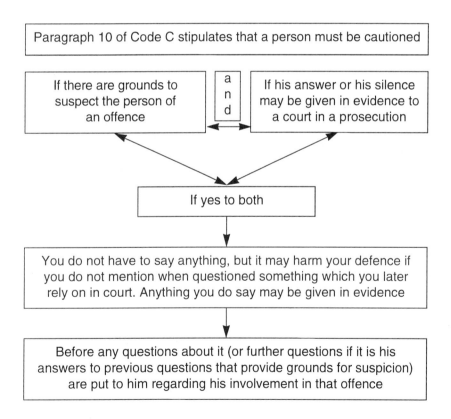

**Figure 5.1** Cautioning under paragraph 10 of Code C (after Bennett, Halford & Steward, *Environmental Health*, Vol 99, No. 9, 1991).

1.   Section 38(3) of the 1994 Act stipulates that a person shall not have a case to answer or be convicted of an offence solely on an inference drawn from a failure such as is mentioned in s.34. Thus silence by itself cannot be taken to establish a prima facie case against an accused person. What is permissible is for the court to draw a negative inference, in certain circumstances, when an accused does not speak. It is not the case that silence at any time can lead to the 'proper inference'.

2.   Section 34 only applies to silence when a person is being questioned under caution; thus silence at any time prior to this and most probably after this cannot be considered.

3.   Any inference that may be drawn from silence is further limited. Only when the failure is to mention facts which are subsequently relied on as part of a defence in court can the inference be drawn. Silence on any other matter is not therefore relevant.

4.   Facts that fail to be mentioned must have been such that in the circumstances existing at the time the accused could reasonably have been expected to mention. Facts which a company officer does not mention, but which he could not reasonably be expected to have known at the time of questioning, cannot give rise to a negative inference.

5.   Finally a court may only draw such inferences as appear proper. It may be that the proper inference in some cases is to recognise that silence was not borne of a bad motive. A company employee may remain silent because of misplaced loyalty or have a greater fear of losing his job than of a criminal prosecution. Of course in some instances an individual may remain silent on legal advice.

On the face of it, it might appear that the provisions of s.34 may permit a negative inference to be drawn when an Environmental Health Officer uses investigative powers other than those under the Police and Criminal Evidence Act 1984. For example s.20(2)(j) of the Health and Safety at Work etc. Act 1974 gives a power to a Environmental Health Officer to question people. However admissibility of any information so gained as evidence is limited by s.20(7) which stipulates that no answer given by a person in pursuance of a requirement imposed under s.20(2)(j) shall be admissible in evidence against that person or the husband or wife of that person in any proceedings. The limitation imposed by subsection 7 will operate such as to mean that any statement obtained from a person using the power conferred by s. 20(2)(j) will not be admissible as evidence against that person. The question thus arises if the inference to be drawn from silence now affects this. The answer lies in s.38(5) of the Criminal Justice and Public Order Act 1994. This provides that nothing in s.34 prejudices the operation of any enactment which provides (in whatever words) that any answer or evidence given by a person in specified circumstances shall not be admissible in evidence against him or some other person in any proceedings or class of proceedings (however described), and whether civil or criminal. Thus if using s.20 of the Health and Safety at Work etc. Act 1974 it is not possible to use the provisions of s.34 of the 1994 Act to draw any negative inference. If it is intended or even anticipated to institute proceedings under the Health and Safety at Work etc. Act 1974 and the Environmental Health Officer intends to put forward any admissions made, or seeks to raise any negative inference from silence, he must use the Police and Criminal Evidence Act 1984.

### 5.5.3 DOCUMENTATION

A record must be made when a caution is given under s.34. This must be either in the Environmental Health Officer's notebook or in the interview record as appropriate. It is always good practice to confirm that the person understands the caution given. If the person cautioned appears not to understand the caution then it is up to the officer to explain its meaning in his own words. It is to be expected that there may on occasion be a break in questioning after cautioning. When there is a break in questioning under caution any interviewing officer must be careful to ensure that the person being questioned is aware that, despite the break, he remains under caution. Where there is any doubt the caution should be given again in full at the resumption of the interview. If an interviewee does not speak English, or has difficulty in understanding English, then an interpreter should be present. If an interpreter is to be used the interviewing officer should ensure that the interpreter makes a note at the time of the interview in the language of the person being interviewed for use should he subsequently be called to give evidence. The interpreter must also certify its accuracy. If a juvenile or a person who is mentally disordered or mentally handicapped is cautioned in the absence of the appropriate adult, the caution must be repeated in the adult's presence. The person being interviewed should be given an opportunity to read the note or have it read to him and sign it as correct or indicate where he considers it incorrect.

### 5.5.4 INTERVIEWS

In the words of Lord Justice Evans in *R v Menard* (1994), questioning is the hallmark of an interview. Under the previous version of the Codes of Practice the distinction between questioning and interview was clear. In the former version of Code C an interview was the questioning of a person regarding his involvement or suspected involvement in a criminal offence; questioning a person only to obtain information or explanation of facts in the ordinary course of an officer's duties did not amount to an interview. An interview for the purposes of the latest version of the code is the questioning of a person regarding his involvement or suspected involvement in a criminal offence or offences which is required to be carried out under caution. As a caution need only be given when there are grounds for suspecting an offence, it would again appear that the questioning of a person only to obtain information or his explanation of the facts or in the ordinary course of the officer's duties does not constitute an interview. An interview for the purposes of the codes is questioning under caution. The difference is vital since the two forms of interaction are regulated in completely different ways. An interview is required to be recorded in accordance with the provisions of the code, questioning is not. Questioning is often essential in establishing whether an offence has been committed, whether a caution should be administered and an interview follow. To overcome problems of admissibility Environmental Health Officers should be aware of the difference but acknowledge how one might lead to the other. While an attempt to disguise as an information gathering exercise what is in reality an interview will not be tolerated, where a suspect volunteers information, not necessarily limited to his alleged offence, and the officer does no more than make notes for their own records, that will not amount to an interview (*R v Menard* (1994)).

In *R v Weerdesteyn* (1994) a customs officer's series of questions, designed to elicit an incriminating response, came within the spirit and letter of an interview. Any note made of the questioning should have been seen by the accused at the time and either approved or a written record made of his disagreement. An accused questioned in this way was to be in full knowledge that he was under no obligation to respond and that any answer given might be used against him. The court found that if a note had been taken of what the customs officer had treated as an interview, despite the accused's lack of knowledge, and the accused had not been shown it when he was still on the premises, nor had any knowledge of it until the criminal proceedings, that untested note constituted a grave breach of the code and evidence obtained in such circumstances, which was central to the point the prosecution had to prove, should have been excluded as seriously incriminating. Increasingly Environmental Health Officers find themselves engaged in the interviewing of suspects in various situations. Not only must the Environmental Health Officer have regard to the provisions of Code C but also to other relevant sources of information. Existing authoritative guidance aims to provide direction on the interviewing of suspects and,

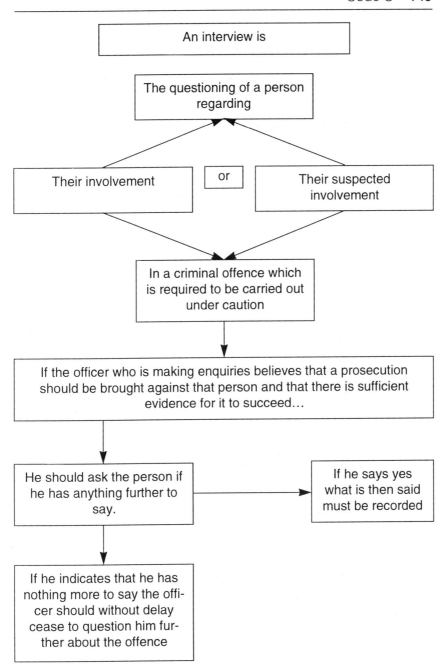

**Figure 5.2** An interview (after Bennett, Halford & Steward, *Environmental Health,* Vol 99, No. 9, 1991)

within that, on the need to observe the provisions of the Police and Criminal Evidence Act 1984 codes. But beyond this Home Office Circular 22/1992 offers general guidance on the broader principles of investigative interviewing. This form of interviewing is wider in scope than the interviewing of suspects addressed in the codes. For these purposes it covers questioning at any stage in an investigation and includes the questioning of not only suspects but witnesses, and victims. The fundamental principles and techniques of interviewing are identified as being common to all these interviews. The central need to acquire accurate, relevant and reliable information when questioning witnesses and victims remains as important as when questioning suspects. The circular, recognising that there are not the same governing rules, provides principles intended to address that omission.

### 5.6 PRINCIPLES OF INVESTIGATIVE INTERVIEWING

The object of any interview is to ascertain the truth, not to prove a case. Though there is no doubt that an interview can assist and add to other evidence to suggest guilt it may also suggest the other alternative. To assist an investigating Environmental Health Officer Circular 22/1992 gives a number of investigative principles applicable to all interviews. These are:

(a) The role of investigative interviewing is to obtain accurate and reliable information from suspects, witnesses or victims in order to discover the truth about matters under investigation.

(b) Investigative interviewing should be approached with an open mind. Information obtained from the person who is being interviewed should always be tested against what the interviewing officer already knows or what can reasonably be established.

(c) When questioning anyone an officer must act fairly in the circumstances of each individual case.

(d) The interviewer is not bound to accept the first answer given. Questioning is not unfair merely because it is persistent.

(e) Even when the right of silence is exercised by a suspect, there is still a right to put questions.

(f) When conducting an interview, officers are free to ask questions in order to establish the truth; except for interviews with child victims of sexual or violent abuse which are to be used in criminal proceedings, they are not constrained by the rules applied to lawyers in court.

(g) Vulnerable people, whether victims, witnesses, or suspects, must be treated with particular consideration at all times.

Circular 22/1992 details the relationship between the principles enunciated and the general rules of evidence. The circular makes it clear that the principles provided are intended to ensure a proper standard of behaviour by any investigating officer taking part in any investigative interview. The principles do not have direct bearing on the rules of evidence. However the circular opines that if observed they should help to provide a natural safeguard against breaching such rules. The purpose of an interview is to obtain from the person concerned his explanation of the facts and not necessarily to obtain an admission. An investigative interview can be particularly useful since it can often indicate likely offences which may be revealed later. For this it may on occasion be appropriate to speak to the person concerned without first cautioning. A caution is only necessary if there is an intention to prosecute on the evidence obtained from the interview. This may not always be the case and it may be that individuals will speak more freely when not speaking under caution. A caveat must be made here: where an interview is conducted without a caution being administered first, there will be very real limits on the admissibility of anything that is said. If answers to exploratory questions give rise to a well founded suspicion that an offence had been committed, what had started out as an enquiry might become an interview. If it does, the proper procedure in accordance with the spirit of the Act and the codes made thereunder is to follow the requirements of the relevant code so far as it is practicable to do so. Accordingly the suspect should be cautioned and a contemporaneous note made of the interview. That applies not only in relation to further questioning but also in relation to material aspects of what had gone before, so that although, for example, a contemporaneous record was no longer a possibility, a record of the earlier questions and answers should be made as soon as practicable. The reason for the lack of a contemporaneous record should be recorded and the suspect given the opportunity to read the record (*R v Park*)

### 5.6.1 INTERVIEW RECORDS

For interviews done under the PACE codes it is essential that an accurate and reliable record is made of each interview with a suspected person, independent of where that interview is conducted. This record must state the following:

1. The place of interview
2. The time it begins and ends
3. The time the record is made (if different)
4. The time of any breaks in interview
5. The names of persons present.

And it must be on a form provided or in the Environmental Health Officer's notebook or in accordance with Code of Practice E on the tape recording of interviews. The record must be made during interview unless that is impracticable or would interfere with the conduct of the interview, in which case it must be made as soon as practicable afterwards. If it is not made during the interview the reason for this must be noted in the officer's notebook. Such a record must be either a verbatim record of what was said or failing this an account which adequately and accurately summarises it. Clearly the verbatim account is the first and best option. All written interview records are required to be timed and signed by the maker. Any refusal by a person to sign must itself be recorded. Based on the requirements of the Police and Criminal Evidence Act 1984 codes a specimen form of interview record sheet is reproduced.

### 5.6.2 CONDUCT OF INTERVIEWS

With regard to the conduct of interviews there are certain matters an Environmental Health Officer must observe. Firstly much of the success of an interview will depend on the preparation for that interview. Any interviewing officer must know all the relevant facts which will assist in their understanding of all elements of the offence. The fuller the knowledge, the better prepared the officer is to carry out the interview, the better the interview will be. Home Office Circular 26/1995 on the Tape Recording of Interviews offers advice on preparing for an interview. It stresses how essential it is that before starting any interview interviewing officers remind themselves of the main elements of the offence(s) they believe to have been committed. This is suggested in order that the officers may be clear, when framing questions and directing the course of the interview, which points it is necessary to be able to prove if a prosecution is to be sustained. To facilitate this all officers should have access to a brief reminder of the main elements, the 'points to prove', of the most common offences. To further assist the officers concerned all available documentary evidence should be on hand. As part of the preparation the Environmental Health Officer must ensure that he or she knows the defences open to the interviewee for the offences under investigation. Adequate preparation can anticipate these defences and ensure the proper questions are asked to ascertain whether such a defence is to be expected and indeed whether it may be viable. Interviews should ideally take place in interview rooms which must be adequately lit, heated and ventilated. Appropriate seating must be provided, an interviewee should not be required to stand. Once the interview is commenced the interviewing officer must identify himself and any other officers present dur-

**Page 1 of.............. Case REF. No..............**
**RECORD OF INTERVIEW**

Date:................................

Location:..................................................

Name and Designation of Officer
conducting interview:

.........................................................

Other Persons present:...........................

| Exhibit No....................... |
| No of Pages................... |
| .......................................... |
| Signature of interviewing officer producing exhibit |

NAME OF INTERVIEWEE

...........................................

OCCUPATION

...........................................

ADDRESS

...........................................

...........................................

Tel No................................
D. o. B. ...............................
Do you wish for a Solicitor to
be present: YES/NO

For Offences by Companies

FULL NAME & ADDRESS OF COMPANY

...........................................................

...........................................................

...........................................................

Are you able to speak on behalf of the
company:
YES/NO

**I understand that I do not have to say anything. But it may harm
my defence if I do not mention when questioned something which
I later rely on in court. I understand that anything I do say may be
given in evidence.**

Signed:................................................Date.....................Time.............

| Person Speaking | |
|---|---|
| | |
| | Contd. |

Times of Breaks............................................................
Time interview commenced: ....................Time interview concluded....................
Time record made...................... Record made by...............................................

**I have read this record of interview and accept it as a true and accurate
record of the interview**
**Signed** ...............................................**Date**...............................................**Time**.............
Others present at the Interview
Signed ...............................................................................................................

**Form 5.2** Record of interview

ing the interview by name. In the interview an officer must not try to obtain answers to questions or elicit a statement oppressively. He must not indicate, except in answer to a direct question, what action will be taken in the event of a person answering questions, making a statement, or refusing to do either. If such a direct question is asked, it is accepted as permissible to inform the person what action is proposed. This is provided that that action is proper and warranted. Although it is probably difficult to envisage, an interviewing officer must not allow intoxicating liquor to be given to any interviewee. Any interviewee who through drink or drugs is unable to appreciate the significance of questions and answers should not be questioned. If interviewing is prolonged there should be breaks from the interview at recognised meal times. There should also be short breaks for refreshment at two-hourly intervals (but these may be delayed in cases of risk to persons or property, unnecessary delay in release from custody, or prejudice to the investigation). If there is any complaint regarding treatment during the interview, which is very unlikely for most interviews by Environmental Health Officers, this must be recorded in the record of interview.

As soon as the Environmental Health Officer who is making enquiries of any person about an offence believes that a prosecution should be brought against that person and that there is sufficient evidence for it to succeed, he should ask the person if he has anything further to say. If the person indicates that he has nothing more to say the officer should without delay cease to question him further about the offence.

A juvenile or a person who is mentally disordered or mentally handicapped, whether suspected or not, is not to be interviewed or asked to provide or sign a written statement in the absence of the appropriate adult unless para. 11.1 of the code or Annex C applies. If a mentally disordered or mentally handicapped person is cautioned in the absence of the appropriate adult, the caution is required to be repeated in the adult's presence.

### 5.6.3 INTERPRETERS

Where an Environmental Health Officer finds that an interpreter is necessary for an interview to be conducted fairly he must ensure that one is available. Code C stipulates that a person must not be interviewed in the absence of a person capable of acting as interpreter if:

(a) he has difficulty in understanding English;
(b) the investigating officer cannot himself speak the person's own language; and
(c) the person wishes an interpreter to be present.

Any interviewing officer must ensure that the interpreter makes a note of the interview, at the time, in the language of the person being interviewed for use in the event of his being called to give evidence. He must also certify its accuracy. The interview should be conducted so as to allow sufficient time for the interpreter to make a note of each question and answer. The interviewee must be given an opportunity to read the interview record or have it read to him and he must sign it as correct or indicate the respects in which he considers it inaccurate. For a tape-recorded interview Code E applies.

## 5.7 CODE E

### 5.7.1 TAPE RECORDING OF INTERVIEWS

Section 60 of the Police and Criminal Evidence Act 1984 provides that:

(1)   It shall be the duty of the Secretary of State –
    (a) to issue a code of practice in connection with the tape-recording of interviews of persons suspected of the commission of criminal offences which are held by police officers at police stations; and
    (b) to make an order requiring the tape-recording of interviews of persons suspected of the commission of criminal offences, or of such descriptions of criminal offences as may be specified in the order, which are so held, in accordance with the code as it has effect for the time being.

Code E deals with the tape recording of interviews with suspects and is best read in conjunction with Home Office circular 76/1988. The circular, when addressing the scope of the code, identifies that a court may exclude as inadmissible evidence obtained in breach of its provisions. In laying down procedures for the police it also identifies best practice in areas which affect other agencies in the criminal justice system and recognises that the tape recording of interviews with suspects will have important implications not only for the police but also for other participants in the criminal justice system. As one of those 'other participants' it is becoming increasingly common for an Environmental Health Officer to undertake a tape-recorded interview. It is therefore important that all officers are conversant with the code and other associated guidance. Circular 76/1988 states the objectives of tape recording, in addition to saving time, to be to provide an accurate and reliable account of police interviews with suspects in police stations. They will provide an effective safeguard for the rights of suspects and should also reduce the likelihood of challenge to the admissibility of prosecution evidence based on these interviews. A record of tape-recorded interview is identified as serving a number of purposes. These are:

- To enable a prosecutor to make informed decisions as to whether the case should proceed
- To ascertain if the proposed charges are appropriate
- To anticipate any possible defences
- To be exhibited to the officer's witness statement and used pursuant to section 9 of the Criminal Justice Act 1967 or section 102 of the Magistrates' Courts Act 1980
- To enable the prosecutor to comply with the rules of advance disclosure
- Where the record of interview is accepted by the defence, to be used for the conduct of the case by the prosecution, the defence, and the court.

Such a record will of course represent what the interviewing officer is likely to say if required to give evidence.

The code requires that tape recordings at police stations shall be used:

- for all interviews with persons charged in respect of an indictable offence. Indictable here includes offences triable either way. This guidance merely repeats the provisions of Schedule 1 of the Interpretation Act 1978 which states that an indictable offence means any offence which, if committed by an adult, is triable on indictment, whether it is exclusively so triable or triable either way;
- for an interview which takes place as a result of a police officer exceptionally putting further questions to a suspect about an indictable offence after he has been charged with or informed he may be prosecuted for that offence;
- for an interview in which an officer wishes to bring to the notice of a person, after he has been charged or informed he may be prosecuted, any written statement made by another person or the content of an interview with another.

In addition where an interview takes place with a person who voluntarily attends at the police (or other investigators) and the officer in question has grounds to believe that person has become a suspect, this being the point at which a caution should be administered, then the interview should be tape recorded. In all cases where an interview is taped it is important to remember that it is the whole of the interview, including the taking and reading back of any interview which should be taped.

### 5.7.2 THE INTERVIEW

Guidance on interviewing contained in the codes envisages that a special interview room should be available. Though there are a number of authorities which do have such rooms it is still a relatively rare resource to be found in Environmental Health departments. Guidance available on the design of such rooms suggests they should be designed to achieve the objective of providing the minimum level of sound proofing consistent with an intelligible, recognisable record of the interview. The elimination of all extraneous background noise is not necessary. An interview room should be equipped for the purpose. Design guidance suggests that there should be no windows, telephones, posters or other distractions. The aim is to have the interviewee's full attention.[1] When using a tape recorder officers should check that the recording machinery is in working order, that they know how to operate it, and that they or any other officer present at the interview can see the time counter on the recording machine. When an interview is to be taped the Police and Criminal Evidence Act 1984 Code of Practice E, at para. 4.1, requires that when any suspect is brought into the interview room the officer should, in the sight of the suspect, unwrap the tapes, load the tape recorder with clean tapes and set it to record. The officer should then speak to the suspect to tell them formally about the recording. The code requires that he say:

(a)  that the interview is being taped
(b)  his name and rank, or for an Environmental Health Officer his designation, and the name and rank of any other officer present
(c)  the name of the interviewee and any other person present; this may be for example his solicitor
(d)  the date, time of commencement and place of the interview, and
(e)  that the suspect will be given a notice about what will happen to the tapes.

The officer is then obliged to caution the suspect in the same terms as mentioned before. At this point any interviewing officer should put to the person being interviewed any significant statement or silence which occurred before taping began and must ask if that person confirms or denies any earlier statement or silence or whether he wishes to add anything. The code defines a 'significant' statement or silence as one which appears capable of being used in evidence.

### 5.7.3 SEALING THE TAPE

Paragraph 4.15 of the code requires that when an interview is concluded the cases of the cassette(s) of the master tape(s) should be sealed, with a master tape label, and signed by the interviewing officer, the interviewee, and any third party present. If the suspect declines to do this the seal should be signed by the appropriate officer. After this the person interviewed should be given a notice which explains how the tape

**Page 1 of.............. Case REF. No..............**
**RECORD OF TAPE RECORDED INTERVIEW**
NAME OF INTERVIEWEE

.................................................................

Date:.......................................................

Location:.................................................

| Exhibit No........................ |
| No of Pages.................... |
| .................................................... |
| Signature of interviewing officer producing exhibit |

Name and Designation of Officer(s) conducting interview:

.................................................................................

Other Persons present: ................................................................

Time interview commenced:........... Time interview concluded: ..............

Duration of Interview: ...............................................................

TAPE REF NO'S ...........................................................................

| Tape Counter times | Person Speaking | Text Capitals for Interviewee |
|---|---|---|
| | | |

Form 5.3 Record of tape recorded interview

recording will be used and how they or their solicitor can arrange to listen to it if a prosecution follows. Circular 76/88 produces a specimen form for this purpose.

### 5.7.4 PREPARATION OF RECORDS OF INTERVIEW

The satisfactory tape recording of an interview involves far more than simply being able to operate the recording device and ask a few questions. It can often be the single most important element in any investigation, accordingly care must be taken in its execution. Home Office Circular 26/1995 gives advice on the satisfactory tape recording of interviews. The guidelines reproduced in this circular are an amended version of those previously contained in Home Office Circular 21/1992 which has now been withdrawn. The most recent circular provides national guidelines on the preparation of records of interview as prescribed by note 5A of Code of Practice E. It recommends that, to ensure a smooth flow to the interview and also that particular parts of it can be easily recorded verbatim afterwards, either the interviewing officer or a colleague should note during the interview the counter times at which anything is said which might later need to be retrieved for verbatim recording or in order to check the accuracy of a third person report.

The circular gives the type of information which is likely to need to be retrieved for either of these purposes. This information includes:

- any admissions and the questions and answers leading up to them
- statements of questions about intent, dishonesty, or possible defences. Matters such as knowledge of key facts, presence at scene of crime on other occasions, or assertion that others were involved should be noted
- aggravating factors which might make the offences be considered more serious
- mitigating factors
- failures to answer questions, whether adequately or at all, which deal with a material part of the allegation.

### 5.7.5 AFTER THE INTERVIEW

After an interview the circular advises that a written record should normally be prepared and this will normally be written in the third person. This record must include, verbatim, all admissions relating to the offence or offences under investigation and the questions and answers which lead up to them. This will include ambiguous admissions (e.g. one of the main elements of the offence may be missing – 'I sold the mouldy food but he knew exactly what he was buying') and qualified admissions (e.g. raising a potential defence – 'He was injured on my machine but it's not possible to get a guard that could have prevented this'). The circular identifies two types of case which are likely to be the subject of a taped interview: complex cases and straightforward cases.

### 5.7.6 COMPLEX CASES

Complex cases are defined as those where:

- the suspect or one of the suspects has been charged with an indictable only offence, or
- the suspect, or one of the suspects has been charged with an either way offence, the circumstances of which make it likely that the magistrates will direct that the case be heard at the Crown Court, or
- the suspect or one of the suspects has been charged with an allegation of assault
- the suspect is likely to deny the offence; this should be assumed in all cases in which the suspect has not admitted the offence and its commission was not witnessed by a [police] officer
- the suspect is under 14 years old.

All other cases are likely to fall in the straightforward category. For complex cases the main salient points must be recorded verbatim. These include statements about intent, dishonesty or defence and knowledge of key elements of the offence. They will also include specifically any questions or answers relating to any other failure by the suspect to answer questions, whether adequately or at all, which deal with a material part of the allegation.

### 5.7.7 STRAIGHTFORWARD CASES

It is likely that for the Environmental Health Officer the majority of tape-recorded interviews will fall into the straightforward category. In these cases, it will usually be appropriate for the record of tape-recorded interviews to be written largely in the third person provided that any admissions and the questions leading up to them are recorded verbatim.

---

**NOTICE TO PERSON**
**WHOSE INTERVIEW HAS BEEN TAPE RECORDED**

This notice explains how the tape recording will be used and the provisions for access to it by you or your solicitor if you are prosecuted.

**Your interview was recorded on ....... tapes.**

**GENERAL**

The interview has been recorded on two tapes. One of these tapes has been sealed in your presence and will be kept securely. It will be treated as an exhibit for the purpose of any criminal proceedings.

The other tape will be a working copy to which the Enforcement Officer(s) and you or your solicitor may have access. Both tapes have protection against tampering.

If you or your solicitor wish to obtain a copy of the tape(s), please send your application to:

............................................................................................................

If you do not have a solicitor now you should consider whether you should seek legal representation. If, however, you do not wish to have solicitor, in the event that you are prosecuted    you    will    be    given    access    to    the    tape    by    applying    to:

............................................................................................................

In all correspondence please quote the following reference number

Ref. No. ...........................    Name of Person(s) interviewed ......................................

**Form 5.4** Notice to person whose interview has been tape recorded

---

**NOTICE TO SOLICITOR**
**TAPE RECORDING OF INTERVIEWS BY ENFORCEMENT OFFICERS**

Name of Defendant..............................................................................................

Tape Ref..............................................................................................................

Reference Number...............................................................................................

The enclosed copies of working copy of...........................tapes relate to the interview with your client. These tapes have/have not* been edited to exclude sensitive material in accordance with the paragraph 6 of Home Office Circular 76/1988.

Should you wish to take issue with any part of the evidence contained on this recording, or require additional material from the recorded interview to be included in the officer's record of interview, you should communicate that fact to

............................................................................................................

at the earliest convenient time.

One of the intentions of tape recording is that any dispute arising out of the conduct of the interview, or of the evidence obtained thereby may be resolved by the defence and prosecution well before the case comes to trial. This will avoid the need for unnecessary adjournments together with the risk of consequential awards of costs being ordered against the defendant or his solicitor.

**Form 5.5** Notice to solicitor

---

### MASTER TAPE SEAL

1. COMPLETE THIS FORM BEFORE AFFIXING TO CASSETTE CASE
2. AFFIX TOP EDGE OF THIS FORM TO TOP OF CASSETTE CASE FACE

---

**DATE** ..................................... **EXHIBIT REFERENCE** ...................................
**PERSON INTERVIEWED**
NAME ..........................................................................................................
SIGNATURE ..................................................................................................
**INTERVIEWING OFFICER**
SIGNATURE ..................................................................................................
NAME AND DESIGNATION .............................................................................

---

**WITNESS**
SIGNATURE ..................................................................................................
NAME AND DESIGNATION .............................................................................

**If the person interviewed refuses to sign above:**
I ........................................................... (the Designated Officer) hereby certify
that this tape was presented to me on ................... (date) at ................. (time)
by ............................................................... (Enforcement officer)

---

SIGNATURE ........................................... DESIGNATION.............................

TIME OF INTERVIEW (START) .............................(END) ...............................
TAPE NO ............... OF ........................ MASTER TAPES USED

**Form 5.6** Master tape seal

## NOTES

1 Further guidance on the creation of an interview room can be found in
'Acoustic Treatment of Rooms – A Step by Step Guide'. Copies are available
from the Information Desk, Home Office Scientific Research and
Development Branch, Woodcock Hill, Sandridge, Herts, AL4 9HQ.

# Statements

# 6

## 6.1 WHAT IS A STATEMENT?

A statement is no more than a documentary record made by a person, which contains details of their direct knowledge and experience of a particular incident. Yet despite the simplicity of this description it is an essential document in the work of an Environmental Health Officer. It is vital because it is via statements that the Environmental Health Officer will both be informed and will inform others of the evidence collected for, and perhaps ultimately to be given in, a court of law.

When investigating an event it may, and indeed will, often be necessary to obtain several statements from a number of individuals before all the facts necessary to establish the offence in a case are given. Consider the Figure 6.1. Imagine for one moment that the Bank 'B' has been robbed. A witness standing at Point 'A' saw two men get out of a red car and rush into the bank; a few minutes later they saw the same men run from the bank and get into the same red car. Witnesses in the bank saw two men produce guns and rob the bank. A few minutes later a witness at Point C saw two men get out of a red car and get into a blue one. He took the number of the blue car. The owner of the blue car and his friend are later arrested. But what witness statements would have been needed to justify this action?

Statements would be needed from witnesses outside the bank. Though they have not seen the actual robbery their evidence is important. It is clearly not an offence to park in front of a bank, nor to enter it at high speed. Therefore their evidence alone is not evidence of a bank robbery. But their evidence might link the red car to the bank and to the men. The witnesses in the bank cannot link the men to the car as they could not have seen it from inside. Their evidence can, though, link the men to the robbery. By taking statements of description and time, the two men carrying out the robbery and the two men entering and leaving

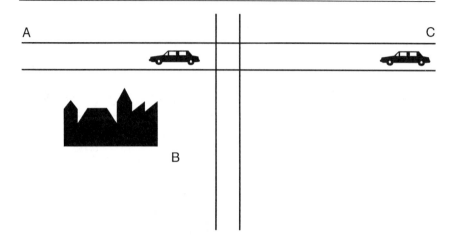

**Figure 6.1** After the robbery of Bank B, statements were taken from witnesses at points A and C.

the bank may be seen to be one and the same. This does not however tie them to the blue car. Yet another statement containing times and descriptions will be needed for this from witness C. No one inside nor outside the bank can have witnessed the men change car several miles away. And again merely to get from one car into another is self evidently not an offence, yet equally again the witness statement is the means by which the men, now in a blue car, are linked to the red car and so to the robbery. Ignoring statements given by police and other investigating officers, it is evident from this very simple example that many statements may be needed from many people to establish all the essential elements of an offence. It is useful to think of statements as the pieces of a jigsaw. A large number of statements may be needed before the picture, that is the offence, can be assembled. Again as a legal competence it is vital that the Environmental Health Officer fully understands all the points to prove in the offence under consideration and, so informed, directs his energies to obtaining witness statements disclosing these.

### 6.2 WITNESS STATEMENTS

A witness statement should contain a record of what an individual directly experienced of an event. Many people will never have given a statement before and the Environmental Health Officer must therefore be able to guide and assist the witness in identifying precisely what it was they saw, heard etc. that is relevant to the investigation. The statement should contain detail of all relevant facts but not include irrelevant matters, which can only serve to obscure the vital information contained in the statement. A brief but comprehensive statement is of more value than a long rambling treatise. It is here that the Environmental Health Officer can be of greatest assistance by ensuring that no relevant fact is overlooked by a witness when giving a statement. If this can be done it will assist the witness in understanding their role in events and the purpose to which their information will be put. It will also greatly assist the Environmental Health Officer in establishing what did happen and therefore what can be proved.

In the statement there should be details of the facts the witness experienced of the event under investigation; these should include details of the day, date, time and location. An EHO, when taking a statement, must ensure that only what the person directly experienced is recorded. Careful questioning may be necessary to ensure that the individual concerned really did experience what they say they did and were not told of it, or some part of it, by someone else. If they were, then of course a statement must be obtained from that other person. Officers should avoid the inclusion of hearsay in statements; this is important, not only evidentially, but also from the point of view of investigation. An officer routinely alert to hearsay will know that he must seek out all witnesses. He will go to all parties capable of giving information on an event; in short his investigation will be more thorough than someone who simply relies on second-hand information. There should be identification, by name if possible, of all people present. This will be useful in helping to ensure all necessary statements are obtained. Anyone not known at the time of the event may, subsequently, be referred to by name by use of the construction; 'I saw a man whom I now know to be F. Bloggs . . .' It is often very much easier for the witness and subsequently for a court if a statement tells its story in a chronological order. This is after all the natural sequence.

An Environmental Health Officer must ensure that any statement given or obtained from a witness contains all details that the witness is able to give. All elements of the offence must be established if the officer intends to take proceedings. Therefore statements from several individuals will often be necessary to establish all the elements of the offence in question. A gap in the evidence because of a failure to obtain a statement or a failure to obtain a complete statement will be fatal to the full understanding of events and will not lead to a complete understanding of who, if anyone, was at fault. Where there is more than one witness in a case it is advisable to produce a witness list for use in any prosecution file. Again format may vary but a typical layout is reproduced in Form 6.1.

The main point when obtaining witness statements is for the Environmental Health Officer to ensure that he guides but does not lead. The Environmental Health Officer must pay particular attention to the requirements of the Police and Criminal Evidence Act 1984 when dealing with juveniles or persons suffering from mental handicap. In such instances, whether suspected or not, a juvenile must not be interviewed or asked to provide or sign a written statement in the absence of the appropriate adult. Appropriate preliminary enquiries, for example asking the age of the interviewee, ought to be able to prevent this from happening.

### 6.3 FORMAT OF A STATEMENT

Witness statements will usually be headed with an annotation: **Criminal Justice Act 1967, s.9, Magistrates Court Act 1980, ss.5A(3)(a) and 5B MC Rules 1981, r.70.**

This may appear on older forms as: Criminal Justice Act 1967, s.9, Magistrates Court Act 1980, s.102, MC Rules 1981, r.70.

Each of these provisions is of relevance to the requirements of a written statement and the use to which it may be put. Section 9 of the Criminal Justice Act 1967 deals with the matter of proof by written statements. The section provides that in any criminal proceedings, other than committal proceedings, a written statement by any person is, if certain conditions are met, admissible as evidence to the same extent as oral evidence and to the same effect.

The conditions which are required to be met for a 's.9 statement' are:

1.  The statement purports to be signed by the person who made it.
2.  The statement contains a declaration by that person to the effect that it is true to the best of his knowledge and belief and that he made it knowing that if it is tendered in evidence he would be liable to prosecution if he wilfully stated in it anything which he knew to be false or did not believe to be true.
3.  Before the hearing at which the statement is tendered in evidence, a copy of the statement is served, by or on behalf of the party proposing to tender it, on each of the other parties to the proceedings; and
4.  None of the other parties or their solicitors, within seven days of the service of the copy statement, serves a notice on the party so proposing objecting to the statement being tendered in evidence under the section.
5.  If a statement is made by a person under the age of 21 (now 18, s.69 Criminal Procedure and Investigations Act 1996) it must give his age.
6.  If a statement is made by a person who cannot read, it must be read to him before he signs it and must be accompanied by a declaration by the person who so read the statement to the effect that it was so read; and
7.  If it refers to any other documents as an exhibit, the copy statement served on any other party to the proceedings under paragraph 3 above must be accompanied by a copy of that document or by such information as may be necessary to enable the party on whom it is served to inspect the document or a copy of it.

**Witness List**

Name and Address of Defendant ........................................................

| |
|---|
| Name<br>Address<br>Occupation<br>Tel No |
| Name<br>Address<br>Occupation<br>Tel No |
| Name<br>Address<br>Occupation<br>Tel No |
| Name<br>Address<br>Occupation<br>Tel No |
| Name<br>Address<br>Occupation<br>Tel No |

**Form 6.1** Witness list

There is nothing in s.9 which prevents the party by whom or on whose behalf a copy of the statement was served from calling the individual giving the statement to give evidence in person nor does it prevent the court requiring such a person to attend; it merely offers a means by which they may not be required to attend. The fact that s.9 statements are not applicable to committal proceedings is attended to by s.5 of the Magistrates Court Act 1980. This section has replaced the former s.102 of the 1980 Act, which is repealed by Schedule 5 to the Criminal Procedure and Investigations Act 1996. This section substantially repeates the provisions found in s.9 of the Criminal Justice Act 1967 but renders permissible the use of statements in committal proceedings.

The Magistrates' Courts Rules 1981, r.70 (as amended by the Magistrates' Courts (Amendment) Rules 1983) requires that written statements to be tendered in evidence under s.5 of the Magistrates Court Act 1980 or s.9 of the Criminal Justice Act 1967 shall be in the prescribed form. The Rules also require that when a copy of such a statement is given to or served on any party to the proceedings, a copy of the statement and of any exhibit which accompanied it shall be given to the clerk of the magistrates court as soon as practicable thereafter, and where a copy of any such statement is given or served by or on behalf of the prosecutor, the accused shall be given notice (by or on behalf of the prosecutor) of his right to object to the statement being tendered in evidence.

It is a requirement that where a written statement, which is to be tendered in evidence under s.5 or s.9, refers to any document or object as an exhibit, that document or object must wherever possible be identified by means of a label or other mark of identification signed by the maker of the statement. The Rules also require that before a magistrates court treats anything referred to as an exhibit in a written statement as an exhibit before the court, that court must be satisfied that the exhibit in question is described sufficiently in the statement for it to be identified. Where a written statement is tendered in evidence under the sections before a magistrates court the name and address of the maker of the statement must be read aloud unless the court otherwise directs.

A statement should so far as possible use the witness's own words to describe events. To hear a statement which clearly uses a form of English totally alien to the way the witness speaks is, to say the least, unconvincing. A court is quite happy to have unusual expressions if they are a true and accurate reflection of what the witness experienced. It is the meaning which is important. The use of terms which fail to convey the facts should be discouraged, but an accurate and understandable, though non-standard, use of English is perfectly adequate. The statement should be written in ink to avoid any suggestion of later alteration. For ease of identification all names should be written in block capitals. This makes any later copying of the statement much easier. If an error is made then the rules with regard to the keeping of a notebook hold good here and the error should be crossed through with a single line to show what the mistake was, as well as the fact that it was a mistake. Each page should be signed by the person giving the statement as must the declaration. Should the author of the statement be unable to read, the statement must be read over to the witness and that fact must be recorded on the statement form. If the Environmental Health Officer has taken the statement, he must clearly identify himself on the statement form. When obtaining a statement it is appropriate to record the availability of potential witnesses. This information can be recorded on the back of the original statement form. It is then a simple matter to examine this and determine dates on which it would be inappropriate to set a trial.

## 6.4 EDITING OF STATEMENTS

### 6.4.1 COMPOSITE STATEMENTS

Where the prosecution proposes to tender written statements in evidence, either under s.5 of the Magistrates' Courts Act 1980 or s.9 of the Criminal Justice Act 1967, it will frequently be not only proper but also necessary for the orderly presentation of the evidence for certain statements to be edited. The editing of statements in such cases is regulated by a 1986 Practice Direction from the Queen's Bench Division. The need to edit will arise either because a witness has made more than one statement, and the contents of those statements can and should conveniently be reduced into a single comprehensive statement, or the statement contains inadmissible, prejudicial or irrelevant material. The Practice Direction is clear on the point that the editing of statements should in all circumstances be done by a legal representative, if any, of the prosecutor and not by an Environmental Health Officer. A composite statement giving the combined effect of two or more earlier statements must be prepared in compliance with the requirements of s.5 of the Magistrates Court Act 1980 or s.9 of the Criminal Justice Act 1967 as appropriate and must of course be signed by the witness.

### 6.4.2 EDITING SINGLE STATEMENTS

The Practice Direction gives two acceptable methods of editing single statements.

1.  By marking copies of the statement in a way which indicates the passages on which the prosecution will not rely. This serves to indicate that the prosecution will not seek to adduce the evidence so marked.

The original signed statement to be tendered to the court should not be marked in any way. According to standard practice the marking on the copy statement should be done by lightly striking out the passages to be edited, so that what appears beneath can still be read, or by bracketing, or by a combination of both. Whenever the striking out/bracketing method is used the following words should appear at the foot of the frontispiece or index to any bundle of copy statements to be tendered: 'The prosecution does not propose to adduce evidence of those passages of the attached copy statements which have been struck out and/or bracketed (nor will it seek to do so at the trial unless a notice of further evidence is served).'

2.  By obtaining a fresh statement, signed by the witness, which omits the offending material.

The guidance given identifies that where a single statement is to be edited the striking out or bracketing method is usually the most appropriate. However it also advises the taking of a fresh statement in the following circumstances:

1.  When an officer's statement contains details of interviews with more suspects than are eventually charged. In this case a fresh statement should be made omitting all details of interview with those not charged except, in so far as it is relevant, for the bald fact that a certain named person was interviewed at a particular time, date and place.

2.  When a suspect is interviewed about more offences than are eventually made the subject of committal charges. A fresh statement should be prepared and signed omitting all questions about the uncharged offence unless they may correctly be taken into consideration or evidence about those offences is admissible on the charges preferred. The Practice Direction suggests that where such omissions are made a phrase such as, 'After referring to some other matters, I then said . . . ', is used to make it clear that part of the interview has been omitted.

3.  If the part of the original statement on which the prosecution is relying is only a small proportion of the whole then a fresh statement should normally be prepared. It does however remain desirable to use the striking out/bracketing method if there is reason to believe that the defence may wish to rely on at least some of those parts which the prosecution does not propose to adduce.

4.  When the passages contain material which the prosecution is entitled to withhold from the defence.

The direction reminds those involved in prosecution that where statements are to be tendered under s.9 of the 1967 Act in the course of summary proceedings there will be a greater need to prepare fresh statements, excluding inadmissible or prejudicial material, rather than using the striking out or bracketing method.

Whenever a fresh statement is taken from a witness a copy of the earlier unedited statement(s) of that witness will be required to be given to the defence in accordance with the Attorney-General's guidelines on the disclosure of unused material unless there are grounds under paragraph 6 of those guidelines for withholding such disclosure. The guidance in question deals with the information to the defence in cases to be tried on indictment and produces five main grounds when the decision not to make disclosure might be rightfully made; these include when a witness might be intimidated, when a statement is believed to be untrue or where the witnesss might change his story.

---

### WITNESS STATEMENT

**(Criminal Justice Act 1967, s.9, Magistrates Court Act 1980, ss.5A(3)(a) and 5B, MC Rules 1981, r.70)**

**Statement of** ................................................................................
**Age (If over 18 enter 'Over 18')** ......................................................

This statement consisting of      page(s) each signed by me is true to the best of my knowledge and belief and I make it knowing that if it is tendered in evidence I shall be liable to prosecution if I have wilfully stated in it anything which I know to be false or do not believe to be true.

Dated the.....................day of.......................................19...................
Signed.........................................................................................

.............................................................................................................
.............................................................................................................
.............................................................................................................
.............................................................................................................
.............................................................................................................
.............................................................................................................
.............................................................................................................
.............................................................................................................
.............................................................................................................
.............................................................................................................
.............................................................................................................
.............................................................................................................
.............................................................................................................
.............................................................................................................
.............................................................................................................
................................................ Signed .....................................
Signature Witnessed by ................................................................

---

**Form 6.2** Witness statement

## WITNESS AVAILABILITY

DELETE WHEN NOT AVAILABLE

Month

| 1 | 2 | 3 | 4 | 5 | 6 | 7 |
|----|----|----|----|----|----|----|
| 8 | 9 | 10 | 11 | 12 | 13 | 14 |
| 15 | 16 | 17 | 18 | 19 | 20 | 21 |
| 22 | 23 | 24 | 25 | 26 | 27 | 28 |
| 29 | 30 | 31 | | | | |

**WITNESS DETAILS**

Address...............................................

.......................................................

.......................................................

Tel No Home................Work...............

Month

| 1 | 2 | 3 | 4 | 5 | 6 | 7 |
|----|----|----|----|----|----|----|
| 8 | 9 | 10 | 11 | 12 | 13 | 14 |
| 15 | 16 | 17 | 18 | 19 | 20 | 21 |
| 22 | 23 | 24 | 25 | 26 | 27 | 28 |
| 29 | 30 | 31 | | | | |

D o B...............................................

Occupation.......................................

Statement taken by...........................

Month

| 1 | 2 | 3 | 4 | 5 | 6 | 7 |
|----|----|----|----|----|----|----|
| 8 | 9 | 10 | 11 | 12 | 13 | 14 |
| 15 | 16 | 17 | 18 | 19 | 20 | 21 |
| 22 | 23 | 24 | 25 | 26 | 27 | 28 |
| 29 | 30 | 31 | | | | |

Month

| 1 | 2 | 3 | 4 | 5 | 6 | 7 |
|----|----|----|----|----|----|----|
| 8 | 9 | 10 | 11 | 12 | 13 | 14 |
| 15 | 16 | 17 | 18 | 19 | 20 | 21 |
| 22 | 23 | 24 | 25 | 26 | 27 | 28 |
| 29 | 30 | 31 | | | | |

Month

| 1 | 2 | 3 | 4 | 5 | 6 | 7 |
|----|----|----|----|----|----|----|
| 8 | 9 | 10 | 11 | 12 | 13 | 14 |
| 15 | 16 | 17 | 18 | 19 | 20 | 21 |
| 22 | 23 | 24 | 25 | 26 | 27 | 28 |
| 29 | 30 | 31 | | | | |

Month

| 1 | 2 | 3 | 4 | 5 | 6 | 7 |
|----|----|----|----|----|----|----|
| 8 | 9 | 10 | 11 | 12 | 13 | 14 |
| 15 | 16 | 17 | 18 | 19 | 20 | 21 |
| 22 | 23 | 24 | 25 | 26 | 27 | 28 |
| 29 | 30 | 31 | | | | |

**Form 6.3** Witness availability

## 6.5 EXHIBITS

Exhibits can assist in establishing the strength or weakness of any witness's evidence. It may be for example that notes used for memory refreshing will become exhibits. If this is so it is not hard to see how, according to the way in which a note was made, it may strengthen or weaken the credibility of a witness. An exhibit must be linked to the statement of the person producing it. This is necessary because that person will usually have to identify and give evidence on it. Exactly how exhibits are linked to statements will depend on the practice of the individual enforcing authority. Usually the officer producing the exhibit will refer to it in his statement and give it a reference number. It is common practice for this to be the officer's initials and a number. Thus an officer in his statement might include the phrases: '. . . . *I took a recording of the noise that I now produce marked AB1. . .' '. . . I wrote a statement under caution that I now produce marked AB1 . . .'*

For any given case there may be more than one exhibit; if this is so these may be referenced as AB2, AB3 etc. It is vital that an officer ensures that there is consistency and care in the recording of exhibits. Carelessness in the labelling and recording of exhibits may result in doubt as to whether they are the items referred to in the statement. This is particularly important where an exhibit is passed from one officer to another. The correct recording of the exhibit number will ensure that an exhibit is accounted for at all times. This information is essential if allegations of subsequent contamination of the exhibit are to be defended. A certificate for exhibit identification will normally be attached to the exhibit; again the precise format for such a certificate is variable although it will generally resemble the example reproduced. Where there is more than one exhibit in a case then it would be advisable to produce an exhibit list for use in any prosecution file..

---

......................................................**Council**

### EXHIBIT IDENTIFICATION

This is the......................................................................
referred to in my statement

| | Signature | Date |
|---|---|---|
| (1)............................................. | .................................. | |
| (2)............................................. | .................................. | |
| (3)............................................. | .................................. | |

Exhibit No ................................... (to be shown at court)

*To be produced by exhibiting officer*

---

**Form 6.4**  Exhibit identification

**EXHIBIT LIST**

Name and Address of Defendant ..............................................................

| ITEM | EXHIBIT REF. NO | PERSON PRODUCING | CURRENT LOCATION OF EXHIBIT |
|------|------|------|------|
|  |  |  |  |
|  |  |  |  |
|  |  |  |  |
|  |  |  |  |
|  |  |  |  |
|  |  |  |  |
|  |  |  |  |
|  |  |  |  |
|  |  |  |  |
|  |  |  |  |
|  |  |  |  |

**Page................ of ................**

Form 6.5 Exhibit list

## 6.6 OTHER STATEMENTS

Some enforcing authorities are in the habit of taking statements using investigative powers found in specific statutory provisions. For example an Environmental Health Officer might take what they term a 'statement' under s.20(1) and 20(2)(j) of the Health and Safety at Work etc. Act 1974. If this is done care must be exercised not to confuse this 'statement' with a s.9 Criminal Justice Act statement. Section 20(2)(j) of the Act gives an inspector the power to require persons, who he has reasonable cause to believe are able to give information relevant to any examination or investigation, to answer such questions as the inspector thinks fit to ask and to sign a declaration of the truth of those answers. This is without doubt a useful investigative tool but is of limited use if it is anticipated that such answers are to be used in evidence. Section 20(7) stipulates that no answer given by a person in pursuance of this requirement is admissible in evidence against that person, or the husband or wife of that person, in any proceedings. If it is intended or anticipated that any admissions made may be later used in criminal proceedings, an Environmental Health Officer would be better advised to adhere to the provisions of the Police and Criminal Evidence Act 1984 and use the s.9 witness statements already described, rather than these alternative 'statements' with their attendant limitations.

## 6.7 PHOTOGRAPHS

There is little doubt that a single photograph can usually present more information than plain text ever could and their use as an aid in investigation is without doubt. In *R v Fowden and White* (1982) it was established that there is no difference, in principle, between a video film, a photograph and a tape recording. Indeed a video recording may be preferable to still photographs and plans and has been held admissible when submitted in preference to both of these (*R v Thomas* (1986)). However, due to the limited availability of video recording equipment still photographs are likely to remain the main means of recording evidence. As photographs would be presented as evidence the question of their admissibility arises. Practically this ought to present little difficulty. Evidence acquired by electronic devices, such as a camera, is admissible and in the words of Sir Joscelyn Simon P in *The Statue of Liberty* [1968]; 'if tape recordings are admissible, it seems equally a photograph ...is admissible.'

Photographs will generally be admissible in evidence if it is proved that they are relevant to the issues in the case and that the prints are untouched and from the original negatives (*R v Maqsud Ali* [1966]). In *R v Stevenson* [1971] it was held that tape recordings had to be proved to be genuine before they were admissible. As there is here no difference

## Health and Safety at Work etc. Act 1974
## Sections 20(1) and 20(2)(j)

I ..............................................FULL NAME AND POSITION IN
ORGANISATION...................................................................................
OF........(ADDRESS)..........................................................................

State that the following is a true and accurate record of the information I gave in answer to questions put to me by ........................................, an authorised officer under the Health and Safety at Work etc. Act 1974.

I declare the above statement to be true

       Signature.............................................

       Date....................................................

Given before me ............................................ as an inspector appointed under Section 19(1) of the Health and Safety at Work etc. Act 1974

*Under the provisions of s.20(1) and s.20(2) of the Health and Safety at Work etc. Act 1974, an inspector appointed under the provisions of s.19(1) has the power to require any person, whom he has reasonable cause to believe to be able to give any information relevant to any examination or investigation, as may in the circumstances be necessary for carrying into effect any of the relevant statutory provisions within the field of responsibility of the enforcing authority which appointed him, to answer (in the absence of persons other than a person nominated by him to be present and any person whom the inspector may allow to be present) such questions as the inspector thinks fit to ask and to sign a declaration of the truth of his answers.*

*Section 20(7) of the Health and Safety at Work etc. Act 1974 stipulates that no answer given by a person in pursuance of a requirement imposed under subsection (2)(j) shall be admissible in evidence against that person or the husband or wife of that person in any proceedings. It is an offence to fail to answer the questions of an inspector or to refuse to sign a declaration as to the truth of the answers given to an inspector, to make a statement which is known to be false or recklessly to make a statement which is false, punishable on summary conviction with a fine not exceeding level 5 on the standard scale.*

**Form 6.6** Health and Safety at Work etc. Act 1974, sections 20(1) and 20(2) (j) between audio tape, video tape and photographs it is therefore necessary that any photograph which is proposed to be introduced into evidence has its provenance verified. This may be achieved by verification on oath, or by statement, by the photographer and, if different, the person who developed the photographs. When photographs are to be used it is normal to obtain a processing certificate from the developers and statements from all persons who have handled the film, negatives and prints. The complete film should be exposed or wound on and the film must be capable of being 'traced' from the purchase of the film to the presentation of prints as evidence. The effect of this is to make available to the court secondary facts or information demonstrating that the exhibit is a true representation of the facts and has not been tampered with or altered in any way. There then can then be no question as to the circumstances (date, time etc.) under which the photographs were taken.

Generally provisions enforced by an Environmental Health Officer will specifically provide for the taking of photographs. For example s.20 of the Health and Safety at Work etc. Act 1974, which deals with the powers of inspectors, confers a power to take photographs of any articles or substances found on the premises. Similarly the Environment Act 1995 permits an inspector to take such measurements and photographs and make such recordings as he considers necessary for the purpose of any examination or investigation. However there may in some statutes be no specific provision for the taking of photographs. The Food Safety Act 1990 makes no explicit provision empowering officers to take photographs, though Code of Practice No. 2 on legal matters suggests that an officer is empowered to inspect anything which may help establish whether or not an offence has been committed. This power, it is claimed, includes the right to take photographic evidence. The code clearly believes there is an implicit power to obtain photographic evidence.

If photographic evidence is to be used there are certain practical points to observe. The officer must first ensure he can use the camera. Increasingly, modern cameras are 'simple but smart' and very forgiving of poor technique. Two shots of each photograph should be taken, which should ensure that any fault on the first frame will not to be repeated on the second. This will not overcome problems with film emulsion or faulty camera shutters but will remove the problem of the finger in the frame. All photographs should be properly presented to the court. They should be of an appropriate size, mounted on stiff card, or presented in the form of a bound booklet. There should be sufficient copies to ensure that if in the magistrates court the witness referring to the photograph has one, each of the magistrates has one and the prosecuting and defence solicitors have one. With the improvement in photocopying technology it is becoming common for colour photocopies to be presented. This has the advantage of relative ease of reproduction and enlargement, however quality must be guaranteed. In any event the originals should be made available to the court. Label the photograph 1, 2, 3, etc. or A, B, C, etc. to aid all parties to identify the photograph in question. It is not appropriate to put arrows on to photographs indicating what is wrong with the photograph. Instead oral testimony should be given as to what the photograph displays. This will prevent problems with hearsay arising. Locating the precise point or points on a photograph can be a problem and there may be a number of such points to be highlighted on a single print. One approach to overcoming this can be to provide a grid for each photograph and to direct the court to the appropriate grid reference. It is important to remember that this grid will not be appropriate for use in all cases and therefore need not be applied to every photograph used.

### 6.7.1 THE PREJUDICIAL BACKGROUND

The defence may object if a photograph shows other offences not currently before the court. Where this is the case, it could be taken to prejudice the offender and may amount to character evidence, which is not normally admissible. The photograph may show not only the offence currently in question before the court but also other offences not put before the court. This may suggest that it is in the character of the individual to commit the type of offence he is charged with because the photograph shows other evidence of similar offences. For example a photograph of a dirty food slicer may in the background show dirty work surfaces. If the individual is not charged with offences relating to the work surfaces then the photograph may be inadmissible. It is in essence saying: look this person is the type of person who would have a dirty food slicer because he also has dirty work surfaces. This is not permissible. It is important therefore that regard is had not simply to what is considered to be the subject of the photograph but to all information contained therein. Officers proposing to use a camera must look carefully at the whole frame of the photograph and ensure that there is nothing which might render the subsequent print inadmissible. The 'prejudicial background' is particularly relevant with regard to video evidence where a camera may track over a wide range and so contain many backgrounds.

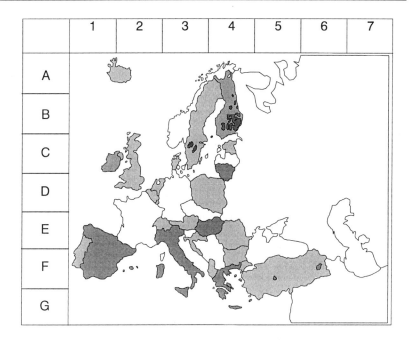

**Figure 6.2** Photographic grid

# Enforcement policy

# 7

## 7.1 ENFORCEMENT STYLES

There can be little doubt of the importance of consistency in enforcement of the law. Inconsistent enforcement in any area of the law may give rise to a lack of public confidence in the enforcing agency and a feeling by those subject to enforcement proceedings that they are being unacceptably victimised. Such inconsistency, where found to exist, is often simply a manifestation of a lack of more formal guidance in the overall regulatory decision-making process. Environmental Health Officers have been criticised in the past for a lack of consistency in their approach to enforcement. Much of this has been proven to be unjustified, nevertheless the perception of inconsistency is itself something which cannot be ignored.

Historically the enforcement of environmental health legislation has been characterised by a relatively low numbers of prosecutions. This has been attributed to an informal regulatory style.[1] A style which in the words of Vogel (1986) is typified by an absence of statutory standards, minimal use of prosecution, a flexible enforcement strategy, considerable administrative discretion, decentralised implementation, close co-operation between regulators and the regulated and restrictions on the ability of non-industry constituents to participate in the regulatory process. This as a style has not always been seen as entirely satisfactory. In 1984 the Tenth Report of the Royal Commission on Environmental Pollution[2] commented that the success of our regulatory authorities needed to be judged on their performance rather than their methods. Yet it is hardly surprising that such a comment could be made. Legislation in the field of environmental health has in the past been perceived as being too often reactive rather than pre-emptory. However change can be identified. For example in the House of Lords it was noted that the volume of environmental law will inevitably increase and increase at a rapid rate.[3] Ball and Bell (1995)[4] note that the British approach is changing quite rapidly, becoming more open, centralised, legalistic and contentious. Recent environmental health statutes have seen provision for increased levels of punishment. For example offenders under the Food Safety Act 1990, Health and Safety at Work etc. Act 1974 and the Environmental Protection Act 1990 all face fines of up to £20,000. Writers now speak seriously about imprisonment for the environmental health offender; indeed in the case of *Bristol City Council v Higgins* (1994) the defendant was jailed for three months for breach of injunction to control noise in domestic premises. Ball and Bell (1995) observe that a number of things appear to be changing in relation to the traditional British approach to regulation and suggest that these changes are taking place in the attitude towards the 'criminality' of the environmental offence.

Yet practitioners express their doubts about the suitability of the criminal process when dealing with environmental health matters and at least one Lord Justice of Appeal has stated that he does: '. . . have reservations as to whether the criminal courts are the appropriate tribunal to determine some of the offences created by environmental legislation'.[5]

Such sentiments are by no means new. There is a long history of ambivalence as to the methods of those charged with protecting the public health. As long ago as 1877 the first Chief Alkali Inspector, R. Angus Smith, recorded his view that:

> It is better to allow some escape occasionally than to bring in a system of suspicion, and to disturb the whole trade by a constant and irritating inspection . . . I . . . work simply . . . by advice and by friendly admonition, and the prosecutions will come in their proper time[6]

This approach to use of prosecution as a weapon of last resort was deeply embedded. So much so that nearly a century later in 1971 the Chief Alkali Inspector in his report stated:

> When the co-operative approach, fails... the time arrives when corrective legal action has to be applied ... Co-operation between all parties is an indispensable part of a successful anti-pollution policy. The Alkali Inspectorate has evolved such a policy over more than one hundred years ... the response has been excellent'.[7]

The Alkali Inspectorate was very influential in the development of our overall systems of environmental management[8] and thus only rarely under this regime could the courts expect to encounter environmental offences. This approach, though understandable, had the effect of creating what McAuslan (1991) identified as a 'secretive closed system of regulation in which great power is vested in officials; a system which has spread into other areas of regulation'.[9] In the recent Report of the Interdepartmental Review Team; Local Government Enforcement, DTI

1994, only in very few authorities was there found to be a clear preference for formal enforcement action (under 10%). The preferred approach was a more flexible one. The survey split the activities into formal action, e.g. prosecutions, notices closure, and alternatives to formal action e.g. time given for compliance, advice. Under 10% of returns showed a marked preference for formal action. Even though Environmental Health Officers have appeared more disposed recently to use prosecution the underlying issue remains. Perhaps it is even heightened, for if more cases are brought before the courts the potential for dissatisfaction with the result and the process rises accordingly. It is this legacy of history which is the cause of many enforcement difficulties that Environmental Health Officers encounter in this area.

Those involved in the enforcement of environmental health law often find themselves uncertain of their role; are they to be a policeman or a friendly advisor? Historically this has resulted in the development of what has been described as an informal regulatory approach. Hutter (1988) identified that many regulatory agencies (not just Environmental Health Officers) had adopted this co-operative approach – an approach which relied on negotiation, bargaining, education and advice to secure compliance.[10] Carson (1970)[11] found that:

> [factory] inspectors do not see themselves as members of an industrial police force primarily concerned with the apprehension and subsequent punishment of offenders. Rather they perceived their function to be that of securing compliance with the standards of safety, health and welfare required and thereby achieving the ends at which the legislation is directed.

This attraction to informal techniques was similarly found by Hawkins (1984)[12] to be displayed by water authorities[13]. Quoting one officer he records that '. . . the objective of the job is not to maximise the income of the exchequer by getting fines. The job is to make the best use we can of the water for the country . . . we get more co-operation if we use prosecution as a last resort.'

As Hutter (1988) points out, the law in books is rarely implemented in a clearcut fashion. It is in this area that the Environmental Health Officers have to determine matters before them, what the law means, whether matters before them are covered by the legislation and then decide on an outcome. It is not enough to simply claim that it is the technical complexity of environmental health cases that causes the problems. The situation is altogether more complex, involving considerations of history, policy and jurisprudence. In this area alone there have been identified a number of interacting factors including:

- the Environmental Health Officers' perception of themselves as educators and advisors rather than as a police force
- the range of penalties available and the perceived low level of penalties imposed
- the exercise of discretion in decision making regarding enforcement
- the accountability of the Environmental Health Officer
- limited resources
- the uncertainty of the criminality of prohibited conduct in this area.

## 7.2 THE CRIMINALITY OF THE ENVIRONMENTAL HEALTH OFFENCE

As has already been explored, there is no substantive definition of what a crime is which can act as an infallible test by which to determine whether any given conduct is criminal. Allied to this is the fact that environmental health law occupies a difficult conceptual position: utilising the criminal law, but operating in an area not entirely suited to its application.

English criminal law is meant to serve a number of purposes and the Court of Appeal has in *R v Sargeant* (1974) laid down four principles to be applied in sentencing which are relevant in understanding the purposes behind the criminal law. These are:

- retribution (retribution being required so that society through its courts can show abhorrence of particular types of crime)
- deterrence (as regards both the offender and others)
- prevention
- rehabilitation.

Such a philosophy, however, is more clearly related and applicable to the concept of 'true criminal activity'. For example it is expected that a person convicted of armed robbery will be punished for their offence, will be given a custodial sentence to protect us from their continued activity and that the sentence may be sufficiently long to act as a deterrent both to reoffending and to those who might be inclined to similar behaviour. The offender in environmental health cases, however, is often perceived to be in a different category to that of the more accepted criminal character.

Ball and Bell (1995) argue that consideration should be given to 'decriminalisation' of whole areas of environmental law so as to allow for a distinction to be drawn between remedies that are properly characterised as administrative and those which are truly criminal and kept for blatant cases of environmental vandalism. With relevant amendments a similar argument could be put forward for much of the general law enforced by Environmental Health Officers. To say this is to restate, with perhaps a different emphasis, what has long been recognised: that the offender in environmental health cases is often perceived to be in a different category to that of the more accepted criminal character. This results in environmental health offences being characterised by Hawkins (1984) as 'economic crimes'. In its report the Robens Committee on Safety and Health at Work [14] noted (at para. 261) that the

> traditional concepts of the criminal law are not readily applicable to the majority of infringements which arise under this type of legislation. Relatively few offences are clear cut, few arise from reckless indifference to the possibility of causing injury, few can be laid without qualification at the door of a particular individual. The typical infringement or combination of infringements arises rather through carelessness, oversight, lack of knowledge or mean, inadequate supervision or sheer inefficiency. In such circumstances the process of prosecution and punishment by the criminal courts is an irrelevancy . . . whatever the value of the threat of prosecution, the actual process of prosecution makes little direct contribution towards this end . . . On the other side of the coin – and this is equally important – in those relatively rare cases where deterrent punishment is clearly called for, the penalties available fall far short of what might be expected to make any real impact, particularly on the larger firms.

Certainly environmental health regulation and enforcement, as it evolves and progresses to meet the challenges now emerging, will increasingly need to acknowledge this dichotomy of perception, and explicitly address it, if the courts are to be assisted when facing these problems.

### 7.2.1 CRIMES AND QUASI CRIMES

The question arises: why should the environmental offender be perceived as being in a different category to that of the more mainstream criminal character? Many of the offences encountered in environmental law are in the nature of regulatory offences. In *Sherras v De Rutzen* [1895], a case under s.16(2) of the Licensing Act 1872, which prohibited a licensed victualler from supplying alcohol to a constable on duty, Wright J stated, at 922, that there

> is a presumption that *mens rea*, an evil intention or knowingly, or a knowledge of the wrongfulness of the act is an essential ingredient in every offence, but that presumption is likely to be displaced either by the words of the statute creating the offence or by the subject matter with which it deals and must be considered (*Nichols v Hall*) . . . the principle classes of exception may be reduced to three . . . one is a class of act which in the language of Lush J in

*Davies v Harvey* are not criminal in any real sense, but acts which in the public interest are prohibited under a penalty.

This was certainly held to include the sale of adulterated food (*Roberts v Egerton*).

The House of Lords, in *Sweet v Parsley* [1970], which concerned the use of drugs on premises, referred to and approved *Sherras v De Rutzen*. Lord Reid stated (at 149) that there were two types of criminal offence:

(a) those which could truly be said to be criminal
(b) those which are not criminal in any real sense, but are acts which in the public interest are prohibited under a penalty.

These have been called regulatory offences or quasi crimes.

In *Alphacell v Woodward* [1972], which dealt with the causation of river pollution under the Rivers (Prevention of Pollution) Act 1951, Lord Salmon placed environmental pollution clearly within the realm of other regulatory offences when (at 848) using the words of Wright J in *Sherras* he said: 'the offences created by the Act of 1951 seem to me to be proto-types of offences which are not criminal in any real sense, but acts which in the public interest are prohibited under a penalty.'

The tendency of the judiciary to characterise many environmental health offences, often because of their absolute nature, as not really criminal reflects a more general view. This is the view that environmental offences are not truly criminal; there are often no obvious victims, and it additionally suggests that they should carry neither the stigma nor perhaps the punishment of true criminal behaviour. Hawkins (1984) identifies this as a moral ambivalence which surrounds regulatory control. He identifies the situation where there is a reluctance to regard breach of regulatory requirements as morally reprehensible. The attendant conduct is often regarded as 'morally neutral'[15] in contrast to those behaviours more normally seen as criminal. This is the well recognised distinction between crimes which are *mala in se* (crimes abhorrent or wrong in themselves) and *mala prohibita* (crimes which result from actions prohibited by law, but which do not arouse deep feelings), or those offences which are, following Curzon (1985), 'technical breaches of the law' and those offences which touch on deep-rooted moral attitudes[16].

Such confusing interpretations then contribute to the situation whereby the courts, when confronted with such an offence, can find it difficult to justify to themselves the imposition of sufficiently harsh deterrent sentences. The courts recognise that culpability may have nothing to do with guilt, and thus, removed from one of the normal reference points of the criminal law, may display a tendency to deal with cases inconsistently. In *R v F & M Dobson Ltd* (1995), a nut brittle sweet was found to be contaminated with a Stanley knife blade and the court was here concerned with the strict liability offence under s.8 of the Food Safety Act 1990. The Court of Appeal found that there were no cases in the Crown Court nor were there any guidelines on the level of sentencing. It was difficult for guidelines to be set because the circumstances in which such cases occurred were, unfortunately, infinitely varied. Their Lordships accepted that culpability had to be an important factor in what was the appropriate penalty to impose but they also considered that deterrence was a major factor. Lord Taylor stated that:

> It was important when the court had to consider such cases, the idea should not go out that any company, providing they had a good record hitherto, could ignore one bite of a Stanley knife blade or some other unwholesome object as something which the court would overlook on the first occasion.

### 7.3 ENFORCEMENT POLICIES

In the absence of formal guidance as to how to deal with such matters the court in *Dobson* felt it must have regard to more informal guidance – informal guidance which could and perhaps should be produced, or at least contributed to, by Environmental Health Officers.

One means of assisting the courts in assessing the seriousness of the offences before them, and of providing the missing reference points, is for Environmental Health Officers to accept explicitly that they have a range of regulatory options available to them – options which allow them to deal with the unwitting offender and the most calculating wrongdoer. However the use of such options, if they are to be of maximum effect, needs to be grounded on a thought-out and publicised body of principles: principles which would guide and direct all enforcement agencies in the correct and appropriate use of the full range of powers available; which would form a body of guidance designed for and aimed directly at those who are engaged in enforcement, and which could, through publication, provide those missing reference points for the courts. If established, such principles could ensure that all parties concerned (enforcers, courts and those subject to regulation) appreciate the seriousness, or not, of the matters under scrutiny. Further this may promote one of the most sought after prizes in the criminal law, consistency. One means of achieving this consistency would be for all Environmental Health departments to develop (as indeed some have) and use a code of practice similar to that of the Crown Prosecution Service. As for the Crown Prosecution Service this could aim to be:

> ... a public declaration of the principles upon which [Environmental Health Officers] will exercise their functions. Having as its ... pur-

pose [the] promot[ion] [of] efficient and consistent decision-making so as to develop and thereafter maintain public confidence in the [Environmental Health Department's] performance of its duties.

Even now only a minority of local authorities have fully developed enforcement policies and even fewer of them are published. This was the finding of the Audit Commission (1991)[17] who found that only 43% of local authorities had a departmental policy for food hygiene and safety law enforcement. Over 90% of respondents to a DTI business survey did not know whether their local authority had an enforcement policy, 3% knew there was one in their local authority, a further 4% said there was no enforcement policy whatsoever in their authority area.[18] The lack of visibility of such policies inevitably raises questions of accountability and effectiveness. Yet it is possible to formulate policies which have within them the means by which effectiveness may be measured. Rowan-Robinson and Ross (1994)[19] have developed a framework which will allow some measurement of effectiveness. They have suggested a nine-point approach as an aid to the measurement of the effectiveness of enforcement. They suggest regard should be had to the following:

1. The clarity of the objectives of the legislation
2. The measurement of unlawful conduct
3. The character of the enforcement agency
4. The resources devoted to enforcement
5. The objectives of enforcement
6. The character of the deviant population
7. The organisation of the enforcement agency
8. External dependency relationships
9. Sanctions for a breach of control.

It has been suggested by Rowan-Robinson and Ross (1994) that the most important institutional pressure which structures the way in which enforcement agencies seek to obtain legislative goals is the agency's own policy on enforcement. They identify that 'the development of a formal "top down" policy compels an agency to address the way in which enforcement practice may accommodate whatever constraints are imposed and still contribute in the most effective way to policy implementation.' In the absence of a 'top down' approach enforcement policy will emerge from the 'bottom up' as the sum of day to day practice by officers. Policy is then formulated, rather than mediated, by inspectors on a case by case basis. This is the least desirable scenario, creating as it does a policy which is opaque, untargeted and unlikely to be commensurate.

## 7.4 DISCRETION

For all enforcing authorities it is not, and never has been, simply a matter of whether or not to prosecute. The imposition of strict liability should not be taken as requiring a prosecution in every case. In *Smedleys Ltd v Breed* [1974], a caterpillar similar in colour, size, density and weight to a pea was found in one of a run of 3.5m cans of peas produced in a season of seven weeks. Viscount Dilhorne asked (at p.855):

> In those circumstances what useful purpose was served by the prosecution of the appellants. Why despite full disclosure made by the appellants was one instituted.

> It may have been the view that in every case where an offence was known or suspected, it was the duty of a food and drugs authority to institute a prosecution, that if evidence sufficed a prosecution should automatically be started . . . I do not find anything in the Act imposing on [an authority] a duty to prosecute automatically whenever an offence was known or suspected and I cannot believe that they should not consider whether the general interests of consumers were likely to be affected when deciding whether or not to institute proceedings . . . The exercise by food and drugs authorities of discretion in the institution of criminal proceedings and the omission to do so where they consider a prosecution will serve no useful purpose is no more the exercise of a dispensing power than the omission of the law officer, the Director of Public Prosecutions and the police to prosecute for an offence. I have never heard it suggested that the failure of the police to prosecute for every traffic offence which comes to their notice is an exercise by them of a dispensing power. No duty is imposed on them to prosecute in every single case and although this Act imposes on the food and drugs authorities the duty of prosecuting for offences under section 2 [Food and Drugs Act 1955] it does not say – and I would find it surprising if it had – that they must prosecute in every case without regard to whether the public interest will be served by a prosecution.

A prosecution where no useful purpose was served was and remains unnecessary. This recognition that not every violation must result in a prosecution gives what must surely be the starting point for any policy on prosecution: discretion. It was Hawkins (1984) who recognised that law in this area may be enforced by compulsion and coercion, or by conciliation and compromise. Existing studies of regulatory agencies have emphasised that enforcement should not be simply equated with prosecution. The term should be used to accommodate a 'much wider concept defining enforcement as the whole process of compelling observance with some broadly perceived objectives of the law' (Hutter 1988).

It is an accepted practice in the enforcement of the criminal law that those charged with enforcement should be given an accompanying discretion as to whether to exercise their enforcement powers or not. As Hawkins (1984) observes: 'The high discretion each officer enjoys is operationally efficient; he is the only one who really knows the dischargers their problems and their negotiating styles.' In *Smedleys Ltd v Breed* [1974] Viscount Dilhorne noted that in 1951 the question was raised as to whether it was a basic principle of the rule of law that the operation of the law is automatic where an offence is known or suspected. He noted that the then Attorney-General, Sir Hartley Shawcross, said that it had never been the rule in this country that criminal offences must automatically be the subject of prosecution and he (the Attorney-General) quoted Lord Simon who in 1925 had said:

> there is no greater nonsense talked about the Attorney-General's duty than the suggestion that in all cases the Attorney-General ought to decide to prosecute merely because he thinks there is what the lawyers call a case. It is not true and no one who has held the office of Attorney-General supposes it is.

Environmental Health Officers are not Attorney-Generals but are surely under a similar duty. In deciding whether or not to take formal action should not the Environmental Health Officer have regard to the general interests of the public? In the area of health and safety at work he may, borrowing from the Health and Safety Executive's approach to enforcement, have as his primary concern the prevention of accidents and ill health, using prosecution as a tool to draw attention to the need for compliance and the maintenance of good standards. In food safety he may seek to ensure that food intended for human consumption is produced, handled, stored and sold in conditions which ensure it is safe and without risk to the health and safety of the public. With appropriate amendments this approach could be adjusted to offer an enforcement policy for all areas of the work of the Environmental Health Officer and would ensure that the level of enforcement is proportionate to the risk to the public involved. Indeed a guided and managed practice of selective enforcement should be taken as an indication of a robust organisation being able to administer its operations in a mature and defensible way. The CPS code recognises that: 'The judicious use of the discretion, based on clear principles, can better serve justice, the interests of the public and the interests of the offender, than the rigid application of the letter of the law.' If it is to do this, however, such an exercise of discretion must be correctly guided and managed. The Crown Prosecution Service recognises that the misuse of discretionary powers can have severe consequences not only for those suspected of crime but also for the public at large and the reputation of justice.

## 7.5 CODES OF GUIDANCE

If legal proceedings are to be instituted it is important that such a decision is not taken lightly or without any reference to policy guidance. To instigate a prosecution against an individual or a company is a serious matter. The consequences for anyone who is the subject of proceedings, even for a relatively minor matter, can be serious. That is even without considering the position of witnesses and the stress which may be imposed upon them. It is important that proceedings are instigated on a fair and consistent basis derived from a mature enforcement policy.

Codes of Practice made under the Food Safety Act 1990 offer one model for the form such guidance may take. The codes recommend that before deciding whether a prosecution should be taken food authorities consider a number of factors, which may include:

(a) the seriousness of the alleged offence;
(b) the previous history of the party concerned;
(c) the likelihood of the defendant being able to establish a due diligence defence;
(d) the availability of any important witnesses and their willingness to cooperate;
(e) the willingness of the party to prevent a recurrence of the problem;
(f) the probable public benefit of a prosecution and the importance of the case;
(g) whether other action, such as issuing a formal caution would be more appropriate or effective;
(h) any explanation offered by the affected company.

This is not, however, the only guidance which Environmental Health Officers can draw on. Both the Health and Safety Executive/Local Authority Liaison Committee (HELA) and the former National Rivers Authority (NRA) adopted policies on enforcement.[20] Although both for their own reasons declined to make the policies fully public, they recognised how important it was to deal consistently with matters. The NRA had a four-point classification system for pollution incidents, aimed at guiding their approach to prosecution.

Pollution incidents were categorised as follows:

| | | |
|---|---|---|
| Major | – | Category 1 |
| Significant | – | Category 2 |
| Minor | – | Category 3 |
| Unsubstantiated | – | Category 4 |

When there was a Category 1 incident where pollution had occurred, where the identity of the discharger was known and sufficient evidence of pollution was available then the National Rivers Authority would have expected a prosecution to follow. In a Category 2 case and where the identity of the discharger was known then the action taken was to be either prosecution, a formal caution, or a letter of warning. The choice was dependent on consideration of all relevant factors. Category 3 incidents were likely to give rise to warning letters if the identify of the discharger was known. A formal caution or prosecution might have been considered where there had been a long record of incidents or the pollution was caused by a deliberate discharge. Category 4 incidents were those which remained incapable of being substantiated. To assist officers in making consistent decisions the National Rivers Authority gave guidance on how to interpret its policy, giving criteria to be used to determine the appropriate action for a category of incident. Though applicable to incidents relating to water pollution it is not hard to see how such an approach may be used to inform the development of enforcement policies for Environmental Health Officers.

**Table 7.1** Factors to consider in deciding action to be taken for Category 2 incidents

| Prosecution or Caution | Warning Letter |
|---|---|
| 1. Discharge deliberate or avoidable | Discharge not foreseeable |
| 2. High risk to abstractions | No risk to abstractions |
| 3. Obvious fish kill | Minor or no fish kill |
| 4. Amenity affected | Amenity not affected |
| 5. Other significant effect on water use (e.g. unfit for stock watering) | Little other impact |
| 6. Negligence | Accidental spillage/discharge |
| 7. Previous history of pollution or breach of consent conditions at site or by polluter | No previous history or evidence of chronic pollution |
| 8. Poor operational management | Good operational management |
| 9. Non-weather related | Weather related |
| 10. No precautions taken | Precautions taken though ineffective |
| 11. Little or no post-incident remedial work | Post-incident remedial work good |
| 12. Consent conditions significantly breached | Consent conditions marginally breached |
| 13. NRA not informed or informed after delay | NRA informed promptly |
| 14. Site security poor | Site security good |
| 15. Pollution prevention advice or literature given | Little or no previous contact |
| 16. Little or no co-operation | Discharger co-operated fully |
| 17. Large number of public complaints | Few or no public complaints |
| 18. Considerable media interest | Little or no media interest |

In the area of health and safety at work the Health and Safety Commission in their 1992 annual report described for the first time the Health and Safety Executive's approach to enforcement. This report made it clear that prosecution is only one element of a strategy which is primarily preventative. The Health and Safety Executive will use prosecution as a tool to draw attention to the need for compliance and the maintenance of good standards. Inspectors investigating breaches of the law will consider their potential to cause harm as well as any harm actually caused. The Health and Safety Commission has now developed an enforcement policy[21] which has embraced and developed this to incorporate the principles of proportionality, consistency, transparency and targeting. Here the Commission defines proportionality as meaning that enforcement action taken by an enforcing authority is to be proportionate to the seriousness of the breach and the risks to health and safety. Consistency relates to the enforcement practice i.e. adopting a similar approach to the options of enforcement rather than ensuring uniformity. Transparency of the arrangements is the extent to which those subject to regulation and the public are clear about what is expected of them and what they can expect of the enforcers. Targeting means enforcement action properly targeted on those who are responsible for the risk and on those whose activities give rise to the risks which are the most serious or least well controlled.

The approach identified by the Health and Safety Executive is echoed by that adopted by the Environment Agency. The Agency has published in its 'Enforcement Policy Statement' the principles that it will adopt for its enforcement activities. When undertaking such activities it will take into account the principal aim of the Agency under s.4(1) of the Environment Act 1995, the Government's Code of Practice for enforcement agencies and the requirements of Schedule 1 of the Deregulation and Contracting Out Act 1994 which advocates codes of practice for regulatory agencies.

The code sets out four basic principles which should inform the enforcement of environmental protection law. As for the Health and Safety Executive these are: Proportionality, Consistency, Targeting of action and Transparency.

For the Agency proportionality means relating enforcement action to the risks and costs. However, the code states that there are some risks so serious that they may not be permitted irrespective of economic and other consequences while at the other end of the scale some risks may be so inconsequential that it would not be worth further expenditure.

For the Agency consistency of approach is again not to mean mere 'uniformity' but adopting a similar approach in similar circumstances. However the Agency recognises that consistency may not be easy because of the number of variables officers may face, which include:

- The degree of pollution
- The attitude and actions of management
- The history of the pollution incidents.

Nevertheless it recognises that decisions on enforcement action are a matter of sound professional judgement and the Agency through its officers should exercise discretion.

In operating transparently officers of the Agency are expected to explain why they intend to take enforcement action and are also required to distinguish compulsory requirements from those which are a matter of guidance and therefore merely desirable.

Targeting for the Agency is a matter of ensuring that inspection and other action is primarily directed towards those whose activities give rise to the most serious environmental damage.

So far as prosecution policy is concerned the Agency will use discretion in deciding whether to initiate proceedings. It is mindful that other approaches to enforcement may prove more effective than prosecution. However, it does recognise that where circumstances warrant it prosecution without prior warning may be appropriate.

The Agency will consider prosecution:

- where it is appropriate in the circumstances as a way to draw attention to the need for compliance and the maintenance of standards, especially where prosecution would be a normal expectation or where deterrence may be a consideration;
- where there has been potential for considerable environmental harm arising from the breach;
- where the gravity of offence, taking into consideration the offender's record, warrants it.

Currently the Health and Safety Executive may seek prosecution if the breach carries significant potential for harm, regardless of whether it caused an injury. In deciding whether to prosecute, the Health and Safety Executive, and those local authorities adopting the HELA guidance, will also consider:

- the gravity of the offence;
- the general record and approach of the offender;
- whether it is desirable to be seen to produce some public effect, including the need to ensure remedial action and, through the punishment of offenders, to deter others from similar failures to comply with the law;
- whether the evidence available provides a realistic prospect of conviction.

In these respects the Health and Safety Executive and the Environment Agency are clearly guided by the Code for Crown Prosecutors published by the Crown Prosecution Service.

## 7.6 THE CROWN PROSECUTION SERVICE

The Crown Prosecution Service in its Code for Crown Prosecutors offers useful guidance on when proceedings are appropriate. The code, which is issued under s.10 of the Prosecution of Offences Act 1985 and is a public document, is based on principles which have hitherto guided all who prosecute on behalf of the public. These principles obviously apply with no less vigour to Environmental Health Officers; indeed the code makes it clear that it contains information which is important to all who work in the criminal justice system. The CPS stipulate two main tests to be considered when prosecuting: The Evidential Sufficiency Tests and The Public Interest Test.

### 7.6.1 EVIDENTIAL SUFFICIENCY

The guidance states that a prosecution should not be started or continued unless the prosecutor is satisfied that there is admissible, substantial and reliable evidence that a criminal offence known to the law has been committed by an identifiable person or persons. The test to be applied is whether there is a realistic prospect of a conviction. A realistic prospect of conviction is identified as an objective test and describes the circumstances when a jury or magistrates, properly directed in accordance with the law and aware of all relevant facts, are more likely than not to convict the defendant of the charge alleged.

There are certain matters that a prosecutor is expected to consider when evaluating evidence. These include:

1. The requirements of the Police and Criminal Evidence Act 1984.
2. Any doubt on any admissions by the accused due to age, intelligence or apparent understanding of the accused.
3. Reliability of witnesses. Has a witness a motive for telling less than the whole truth? Might the defence attack his credibility? Are all the necessary witnesses available and competent to give evidence?

### 7.6.2 THE PUBLIC INTEREST CRITERIA

If the evidential requirements are met the CPS then propose that a prosecutor must consider whether the public interest requires a prosecution. In this the CPS is guided by the view, already mentioned, expressed by Sir Hartley (later Lord) Shawcross when he was Attorney General:

It has never been the rule in this Country – I hope it never will be – that suspected criminal offences must automatically be the subject of prosecution. Indeed the very first Regulations under which the Director of Public Prosecutions worked provided that he should . . . prosecute 'wherever it appears that the offence or the circumstances of its commission is or are of such a character that a prosecution in respect thereof is required in the public interest' That is still the dominant consideration . . . the effect which the prosecution, successful or unsuccessful as the case may be, would have upon public morale and order, and with any other considerations affecting public policy.

### 7.6.3 THE DECISION NOT TO PROSECUTE

The factors which may lead to a decision not to prosecute will of course vary from case to case. However broadly the CPS recognises that the graver the offence, the less the likelihood that the public interest will allow of a disposal less than prosecution. The code gives guidance on a number of matters in the decision to prosecute. Such matters include the following.

#### 7.6.3.1 The likely penalty

When the circumstances of an offence are not particularly serious, and a court would be likely to impose a purely nominal penalty, prosecutors should carefully consider whether the public interest would be better served by a prosecution or some other form of disposal.

#### 7.6.3.2 The offence was committed as a result of a genuine mistake

If the offence results from a genuine mistake or misunderstanding these may be factors against prosecution but must be balanced against the seriousness of the offence.

#### 7.6.3.3 If the loss or harm caused can be described as minor and was the result of a single incident

This would be particularly relevant if the harm was the result of a single misjudgement.

#### 7.6.3.4 Staleness

Prosecutors are advised to be careful of prosecuting if there has been a long delay between the offence and the probable date of trial, unless the offence is serious or the delay has in part been caused by the defendant or the offence has only recently come to light or the nature of the offence has required a long investigation.

Generally, the graver the allegation, the lesser the significance that will be attached to the element of staleness.

### 7.6.3.5 The effect on the victim's health

A prosecution is less likely to be needed if it is likely to have a very bad effect on the victim's physical or mental health; this is, of course, always bearing in mind the seriousness of the offence.

### 7.6.3.6 Old age and infirmity

The older or more infirm the offender, the more reluctant the Crown Prosecutor may be to prosecute unless the offence is serious or there is a real possibility of the offence being repeated.

It may also be necessary to consider whether the accused is likely to be fit enough to stand his trial and regard should be had to any medical reports. The Crown Prosecution Service must balance the desirability of diverting a defendant who is suffering from significant ill health with the need to safeguard the public.

### 7.6.3.7 The defendant has put right any loss or harm

In certain cases it will be appropriate for the prosecutor to have regard to the actions of a defendant in voluntarily compensating a victim. It is important however that this is not seen as an offender simply being able to buy their way out of trouble.

### 7.7 ASSESSMENT OF FORMAL ACTION

Each case will be different and, as the Crown Prosecution Service points out, it must never be for them, nor therefore for the Environmental Health Officer, that it becomes simply a matter of adding up the number of factors on each side of the argument. Though each of the foregoing factors are certainly relevant to the Environmental Health Officer when formulating a policy and deciding upon a course of action, the importance of each factor must be decided upon by the Environmental Health Officer on a case by case basis. This is what is, after all, meant by the officer 'mediating the enforcement policy'. It is not possible to state the relative importance of each matter for every conceivable case, there are too many. It is however possible to assist in structuring the decision and thereby to attempt to ensure all relevant factors are considered. To do this there is reproduced a pro-forma 'Formal Action Assessment Guide', designed as an aid to decision making. This cannot hope to fully cover all matters which may be relevant when examining any potential offence and, though it may be a useful aid, it is important for an officer to remember to have regard to all relevant features of the offence in question, particularly those which mark the offence as serious. Samuels (1994)[22] states that the courts should insist that they are told the full extent of the seriousness of any offence. For example they should be told of the quantity of material dumped, the nature and degree of toxicity, the actual and potential polluting effects, the danger to people and the environment and the cost of removing it or controlling it. He also suggests other factors which would be relevant for the courts to be aware of when dealing with an environmental offence. These include commercial gain and the responsibility of the defendant. The former because it is relevant to know what the likely commercial gain the defendant made, or was likely to make, was and the latter to ascertain the degree of responsibility of the defendant in relation to any other people involved. Again to assist in this and accompanying the Action Assessment Guide are two tables. These contain some of the most common aggravating and mitigating features associated with the whole range of potential Environmental Health offences. By looking for and weighing these features an Environmental Health Officer ought to be better able not only to scale his response to the offence but also to identify for the court those features which make the offence particularly serious. It might also prove useful in anticipating those matters which may be put forward in mitigation at any later trial.

## 7.8 THE CAUTIONING OF OFFENDERS

As already recognised an Environmental Health Officer has a number of options when confronted with any given situation. This may range from taking no formal action, service of notice, to prosecution, with seizure of profits and disqualification of directors. Food Safety Code of Practice No 2 'Legal Matters' indicates what factors a food authority should have regard to before deciding whether to prosecute. It also makes it clear however that one viable alternative to prosecution, what might be considered an intermediate option, is the administration of a formal caution under the provisions to be found in Home Office Circular 18/1994. This circular replaced the earlier Circular 59/1990, which was itself a replacement of Home Office Circular 14/1985. The purposes of this circular are to provide guidance on the cautioning of offenders, and in particular –

- to discourage the use of cautions in inappropriate cases, for example for offences which are triable on indictment only;
- to seek greater consistency;
- to promote the better recording of cautions.

## FORMAL ACTION ASSESSMENT GUIDE

Page 1 of.......

| Background Information | |
|---|---|
| **Nature of the Offence**<br>Summarise including details of<br>relevant statutory provision. | |
| **Full Name of Potential**<br>**Defendant/Notice recipient**<br>If a Partnership Name All Partners | |
| **Address**<br>If a Company Give Reg. Office | |
| **Company**<br>Give Details of Nature of<br>Business/Operations | |
| **Location of Events/Occurrence**<br>Give Details if Different from above | |
| **Name Investigating Officer(s)**<br>**for purposes of Criminal**<br>**Procedure and Investigations**<br>**Act 1996**<br>Duration of Investigation:<br>From............to............ | |
| **Offence Assessment**<br>**(see accompanying aggravating/mitigating factors)** | |
| **Mental Element** | |
| **Seriousness of Offence:**<br>Extent/Size of Offence<br>Impact<br>Injury/Health effect<br>Other Factors: see attached | |
| **External Factors Contributing**<br>**to Offence** | |

Form 7.1 Formal action assessment guide

| Offender Assessment (see accompanying aggravating/mitigating factors) | |
|---|---|
| **Any Mitigating Circumstances, Any Explanation** | |
| **Offender factors** <br> Age <br> Ill Health <br> Co-operation <br> Remorse | |
| **Previous History** <br> Convictions <br> Cautions <br> Warning letters <br>         Give details | |
| **Policy and Procedure** | |
| **Departmental Policy on This Type of Offence?** | |
| **Public Benefit of Prosecution Summarise** | |
| **Is Any Other Action Available to the Local Authority? Is any Proposed? Has any Occurred?** | |
| **Police and Criminal Evidence Act 1984: Have all elements been adhered to?** | |
| **Criminal Procedure and Investigations Act 1996 Recording of information? Retention of Material?** | |
| **Reliability of Key Witnesses** | |
| **Willingness of Key Witnesses** | |
| **Is a Statutory Defence** <br> Available <br> Anticipated | |
| **Likely Penalty** | |

Form 7.1 *continued*

**Investigating Officer(s) Remarks**

**Is there a Realistic Prospect of Conviction?**...................................

**Recommendation of the Investigating Officer**

**No Action**

**Informal Action (specify)**

....................................................................................................
....................................................................................................
....................................................................................................

**Formal Action (Give Reason)**..................................................
....................................................................................................

**Notices** ................................................................................
....................................................................................................

**Caution**................................................................................
....................................................................................................
....................................................................................................

**Prosecution** ..........................................................................
....................................................................................................

**Signed**                                              **Date**

Form 7.1 Continued

## FORMAL ACTION ASSESSMENT GUIDE

### AGGRAVATING FEATURES OF ENVIRONMENTAL HEALTH OFFENCES
### Offence Aggravation

| | |
|---|---|
| 1. Actual polluting effects | 10. Potential polluting effects |
| 2. Danger created | 11. Real risk of injury to health |
| 3. Direct injury to human health | 12. Real risk of injury to the environment |
| 4. Direct injury to the environment | |
| 5. High remedial cost | 13. Seriousness of risk |
| 6. Lack of procedural control | 14. Toxic material dumped |
| 7. Large quantity of material dumped | 15. Toxic material released |
| 8. Large quantity of material released | 16. Vulnerable victim |
| 9. Occurrence was over a prolonged period. | |

### Offender Aggravation

| | |
|---|---|
| 1. Commercial gain – consider cost of compliance | 8. High public awareness, serious concern |
| 2. Degree of responsibility of the accused – culpable ignorance/ neglect | 9. No reference to authorisation or permit/licence ever having been held |
| 3. Deliberate, calculated, blatant, flagrant, pre-meditated offence | 10. Planned deception |
| | 11. Previous convictions |
| 4. Failure to set up quality system | 12. Previous warning/notices/ cautions ignored |
| 5. Failure to set up safe system | |
| 6. Gross disregard for authority | 13. Repeated breaches |
| 7. Group act/conspiracy | 14. Sophisticated planning |
| | 15. Worker awareness |

**Form 7.2** Formal action assessment guide: aggravating features of environmental health offences

## FORMAL ACTION ASSESSMENT GUIDE

### MITIGATING FEATURES OF ENVIRONMENTAL HEALTH OFFENCES

#### Offence Mitigation

| | |
|---|---|
| 1. Accidental oversight | 16. Natural events aggravated damage caused |
| 2. Adequate staff supervision | |
| 3. An unanticipated sequence of events | 17. No injury/damage |
| 4. Circumstances could not be anticipated | 18. Offence attributable to genuine mistake |
| 5. Codes of Practice followed | 19. Offence attributable to third party |
| 6. Confusion of responsibility | 20. Professional advice sought and followed |
| 7. Good operating practice followed | |
| 8. Hygiene/Safety/Environment policy produced and in operation | 21. Sampling and analysis done by accredited laboratories |
| 9. Incident had never happened before | 22. Short period |
| 10. Licence held but not covering process under consideration | 23. Single item involved |
| | 24. Staff adequately trained and informed |
| 11. Little risk of injury or damage | |
| 12. Machine failure due to latent defect | 25. Suitable technology used |
| 13. Machinery correctly maintained | |
| 14. Minor obstruction | |
| 15. Momentary lapse | |

#### Offender Mitigation

| | |
|---|---|
| 1. Age and history of defendant | 5. Previous good record |
| 2. Co-operation with investigation | 6. Speedy attempt to remedy defects |
| 3. Impulsive action | 7. Voluntary compensation |
| 4. Indirect effect of conviction or sentence | 8. Willingness to pay clean up costs |

**Form 7.3** Formal action assessment guide: mitigating features of environmental health offences

Though most obviously aimed at police officers the circular is nevertheless appropriate for those involved in Environmental Health enforcement. LACOTS has, for example, recorded that there has been a national
willingness to identify formal cautions as an enforcement option in the
area of food safety. Formal cautioning need not however be limited only
to this area. It can successfully be incorporated into an enforcement policy relevant across the whole range of the work of the Environmental
Health Officer. However to be most effective this enforcement option
must be deployed as part of such a policy and not considered in isola-
tion. The Home Office circular recognises the difficult interface that
exists between informal warnings and formal cautions and between formal cautions and prosecutions. To attempt to use the formal caution outside a framework of guidance would be to run the risk of misusing a
valuable tool. But, properly used, cautioning is to be regarded as an
effective form of disposal

## 7.9 THE DECISION TO CAUTION

The decision to caution is one for the enforcing authority and should not
be merely a matter of routine.

The Home Office circular seeks to promote greater consistency and
the better recording of cautions. It does not and it is not intended to lay
down hard and fast rules as to the use of cautioning. Ultimately the
proper use of discretion is a matter to be guided by a thorough and
refined enforcement policy, having regard to whether the circumstances
are such that the caution is likely to be effective and appropriate to the
offence. The circular identifies that the accurate recording of cautions is
essential in order both to avoid multiple cautioning and to achieve
greater consistency. A formal caution so recorded will then be expected
to influence an authority in its decision whether or not to institute proceedings if the person should offend again. Additionally, and advanta-
geously, an earlier caution may be cited in subsequent court proceedings
for other offences if the person is there found guilty. A form for the
recording of formal cautions based on those used by other agencies is
reproduced to indicate the nature of the information that may need to be
put before a court (form 7.6).

Before a caution is given departmental or other records should be
checked to ascertain if the offender has received any earlier such warnings. It is both possible and permissible for a person or organisation to be
cautioned on more than one occasion, though the policy implications of
this should be acknowledged. The circular is aware that multiple cautioning may bring this option into disrepute and it advises that cautions
should not be administered to an offender in circumstances where there

can be no reasonable expectation that this will curb his offending. Guidance suggests that it is only in the following circumstances that more than one caution should be considered:

- where the subsequent offence is trivial;
- where there has been a sufficient lapse of time since the first caution to suggest that it had some effect.

This is consistent with the identified purpose of formal cautioning which is to deal quickly and simply with less serious offenders, to divert them from the criminal courts and to reduce the likelihood of their reoffending. When an offender is cautioned on the same occasion for more than one offence then the guidance stipulates that this should be counted as his having received one caution only. Recognising that there will be in any enforcement policy the option of simply giving a verbal warning the circular is clear that there is no intention to inhibit this practice. Care should be taken however to ensure that this is not recorded as a formal caution. Unlike such a caution a verbal warning may not be cited in subsequent court proceedings. It should be clearly understood that a formal caution is not a sentence of the court and cannot be made conditional on the completion of a specific task.

Because of the seriousness of the decision to caution it should not be taken lightly. The circular in recognising this stipulates certain conditions which are required to be met before a caution should be given:

1. there must be evidence of the offender's guilt sufficient to give a realistic prospect of conviction;
2. the offender must admit the offence;
3. the offender must understand the significance of the caution and give informed consent to being cautioned.

Those proposing to use this method of disposal are warned to ensure that consent to the caution is not to be sought until it has been decided that cautioning is the correct course. The significance of the caution must be explained. The recipient must understand that a record will be kept of the caution, that the fact of a previous caution may influence the decision whether or not to prosecute if the person or company should offend again, and that it may be cited if the person or company should subsequently be found guilty of an offence by a court. Where the evidence does not meet the required standard the circular is clear that a caution cannot be administered. Nor will it be appropriate where a person does not make a clear and reliable admission of the offence. Cautioning is an alternative to a prosecution and therefore ought not to be administered where a prosecution could not be commenced.

## 7.10 PUBLIC INTEREST PRINCIPLES

In the giving of a caution, as in any prosecution decision, there are public interest considerations which apply and which will inform the decision whether or not to administer a caution. These public interest principles are as described in the Code for Crown Prosecutors, i.e.:

- the nature of the offence
- the likely penalty if the offender were to be convicted by a court
- offender's age and state of health
- previous criminal history
- attitude of offender towards the offence, including practical expressions of regret.

In giving its guidance the circular clearly has in mind the more mainstream criminal activities, particularly when it considers the role and views of the victim when administering a formal caution. Yet there is no reason why this guidance should be viewed as inapplicable to the Environmental Health Officer. It is advised that before a caution is administered it should be viewed as desirable that any victim should be contacted to establish his or her view about the offence and the nature and extent of any harm or loss suffered. These should then be assessed, relative to the victim's circumstances, and consideration should be given as to whether the offender has made any form of reparation or paid compensation. In some, though obviously not all, environmental health offences this would be sound guidance to follow. If a caution is being considered it is little more than common sense and courtesy that its significance should be explained to any victim.

## 7.11 CAUTIONING OFFICERS

The code suggests that police officers administering cautions should be of a certain rank but does consider that it may be appropriate to nominate suitable cautioning officers. The concept of an officer of 'rank' is inappropriate for Environmental Health Officers, though certainly seniority is a more accessible concept.

It is important that the giving or receiving of a caution is not seen as a soft option. This is why all formal cautions must be recorded and thorough records kept. Where a caution is to be issued to a partnership it is important that the records show all recipients. In subsequent proceedings formal cautions may be cited in court if they are relevant to the offence then under consideration. This will occur after a finding of guilt and though the information cannot be relevant to any finding of guilt or innocence, it may have direct bearing on any sentence imposed. Care must be taken to distinguish previous cautions from previous convictions. Though the circular clearly envisages face to face formal cautioning guidance published by LACOTS suggests that there is no bar on the use of a formal cautioning system which relies on acceptance of the caution by letter. Guidance on the suggested method of using the postal system for cautioning is to be found in their document CO 11 94 5 of 26 May 1994.

**C A U T I O N**
**[For use with an individual/partnership]**

**SURNAME:**
**FORENAME(S):**
**ADDRESS:**
**BUSINESS ADDRESS (if appropriate):**
**DATE OF BIRTH:**

**Date of Offence:**

**Place of Offence:**

**Brief Circumstances of Offence**, including details of relevant statutory provision:

**Declaration**
I hereby declare that I admit the offence described above and agree to accept a caution in this case. I understand that a record will be kept of this caution and that it may influence a decision to institute proceedings should I be found to be infringing the law in the future. I further understand that this caution may be cited should I consequently be found guilty of an offence by a Court of Law.

**Signed:**                                                   **Date:**

**Name:**
**Signed:**                                              **(Designation)**
**Date:**                                             **(Name of Authority)**

Form 7.4 Caution (individual)

# C A U T I O N
### [For use with a Corporate Body]

**TRADING NAME:**
**FORENAME(S) OF OFFICER OF COMPANY:**
**SURNAME:**
**DATE OF BIRTH:**
**POSITION WITHIN COMPANY:**
**ADDRESS OF REGISTERED OFFICE (if no Reg. Office give place of business):**

**Date of Offence:**

**Place of Offence:**

**Brief Circumstances of Offence** including details of relevant statutory provision:

**Declaration**
I hereby declare that on behalf of the above mentioned company I admit the offence described above and agree to accept a caution in this case. I understand that a record will be kept of this caution and that it may influence a decision to institute proceedings should the company be found to be infringing the law in the future. I further understand that this caution may be cited should the company consequently be found guilty of an offence by a Court of Law.

**Name:**              **Signature:**              **Date:**
**Position within the company**

**Name:**
**Signed:**                          **(Designation)**
**Date:**                            **(Name of Authority)**

Form 7.5  Caution (corporate)

# PREVIOUS CAUTIONS

**Individual**

Surname...............................................Forename(s)...............................

D.o.B..................................................................................................

Address...............................................................................................

.............................................................................................................

**Company**

Name...................................................................................................

Corporate/
Unincorporate.....................................................................................

For Partnership give names of all partners

.............................................................................................................

.............................................................................................................

.............................................................................................................

Registered Office (If no Reg. office indicate place of business or Principal office of the Partnership)

.............................................................................................................

.............................................................................................................

.............................................................................................................

| Date | Location & Name of Officer Administering Caution | Offence Details (inc. full details of statutory provision contravened) If Company include name and position of person receiving on behalf of company |
|------|--------------------------------------------------|--------------------------------------------------------------------------------------------------------------------------------------------------|
|      |                                                  |                                                                                                                                                  |

Form 7.6 Previous cautions

## PREVIOUS CONVICTIONS

**File Ref...........**                                    **Page 1 of....**

**Individual**

Surname.................................................Forename(s).........................

D.o.B.................................................................................................

Address............................................................................................

................................................................................................................

**Company**

Name..................................................................................................

Corporate/

Unincorporate....................................................................................

For Partnership give names of all partners

................................................................................................................

................................................................................................................

..............................................................

Registered Office (If no Reg. office indicate place of business or Principal office of the Partnership)

................................................................................................................

................................................................................................................

................................................................................................................

Address (for company give address of Reg. office)

................................................................................................................

................................................................................................................

................................................................................................................

| Date | Court | Offence Details (inc. details of statutory provision contravened) | Sentence Imposed |
|------|-------|---------------------------------------------|------------------|
|      |       |                                             |                  |

**Form 7.7** Previous convictions

## 7.12 POWER TO CHARGE FOR ENFORCEMENT ACTION

Perhaps the need for more transparent enforcement is greater now that in certain areas of work provision is starting to be made to allow an authority to charge for enforcement activities undertaken by Environmental Health Officers and others. Section 87 of the Housing Grants, Construction and Regeneration Act 1996 states that a local housing authority may now make such reasonable charges as they consider appropriate, as a means of recovering administrative and other expenses incurred in taking action of the following kinds:

(a) serving a deferred action notice under s.81 or deciding to renew such a notice under s.84;

b) serving a notice under s.189 of the Housing Act 1985 (repair notice in respect of house unfit for human habitation);

(c) serving a notice under s.190 of that Act (repair notice in respect of house in state of disrepair but not unfit for human habitation);

(d) making a closing order under s. 264 of that Act;

(e) making a demolition order under s. 265 of that Act.

The Secretary of State has the power to specify the maximum amount of any charge and in the Housing (Maximum Charge for Enforcement Action) Order 1996 that amount has been specified as £300. Subsections (2) to (4) of section 87 specify the expenses which may be recovered. These are those expenses incurred in:

(a) determining whether to serve, or where appropriate renew, the notice or make the order;

(b) identifying the works to be specified in the notice; and

(c) serving the notice or order.

This ability to charge may indicate one future trend. If so it must be evident to all that arbitrary or unstructured enforcement policies are inconsistent with defensible charging regimes.

## 7.13 BENEFITS OF ENFORCEMENT POLICY

By the adoption of a policy for enforcement, all having an interest in this area will gain. The Environmental Health Officers, by having a defined and identifiable policy, will be able to demonstrate their performance against publicised and considered criteria. Those subject to regulation will be better able to understand the enforcement policies of the regulators, perhaps being assured that enforcement is not the arbitrary activity that it has sometimes appeared to be. The courts, by having access to clear policy guidelines, will be better able to understand the enforcement policies of the relevant agencies. And the public, by the publication of such policies and associated accountability, may have greater confidence in those charged with protecting their environment and their health.

## NOTES

1 Vogel D, 1986, *National Styles of Regulation,* Cornell University Press.

2 Tackling Pollution - Experience and Prospects, Cmnd 9149,1984, HMSO.

3 ENDS 216 p.26, Law Lords Urge Reform for Environmental Cases, House of Lords Debate of 21 January 1994 on the Implementation of EU Environmental Law.

4 Ball S and Bell S, 1995, *Environmental Law,* Third Edition, Blackstone Press, London.

5 Lord Justice Woolf, 4th Annual Garner Lecture, Are the Judiciary Environmentally Myopic (1992), *Journal of Environmental Law,* Vol. 4, No. 1. p.10.

6 Report of the Royal Commission on Noxious Vapours 1878 cited in Clapp, BW, 1994, *An Environmental History of Britain Since the Industrial Revolution,* London, Longman, and Richardson G, Ogus A, Burrows P, 1992, *Policing Pollution; A Study of Regulation and Enforcement,* Oxford, Clarendon Press.

7 Chief Alkali Inspector, Annual Report 1971, 13.

8 McAuslan P, 1991, The Role of Courts and Other Judicial Type Bodies in Environmental Management, *Jnl. of Env Law,* Vol. 3, No 2, 195, 199.

9 Richardson G, Ogus A, Burrows P, 1992, *Policing Pollution; A Study of Regulation and Enforcement,* Oxford, Clarendon Press.

10 Hutter, BM (1988) The Reasonable Arm of the Law: The Law Enforcement Procedures of Environmental Health Officers, *Oxford Socio-Legal Studies,* Clarendon Press.

11 Carson, WG (1970) White collar crimes and the enforcement of factory legislation, *British Jnl. of Criminology,* vol. 10, pp.383-398.

12 Hawkins K. 1984, *Environment and Enforcement; Regulation and the social definition of pollution,* Clarendon Press, p.179.

13 *Ibid.* p.190.

14 Robens Committee (Cmnd. 5034 (1972)).

15 Kadish (1963) Some Observations on the Use of Criminal Sanctions in Enforcing Economic Regulations, U. Chi. L.R. 423.

16 Curzon LB, 1985, *Jurisprudence,* Macdonald & Evans, Plymouth.

17 Audit Commission, 1991, Towards a Healthier Environment, Managing Environmental Health Services.

18 Local Government Enforcement, Report of the Interdepartmental Review Team, DTI, 1994, HMSO, p.35.

19 Rowan-Robinson J, Ross A, 1994; Enforcement of Environmental Regulation in Britain: Strengthening the Link, JPEL March, p.200.

20 ENDS Report 189, p.28; HELA LAC (L) 563, February 1994.

21 Health and Safety at Work etc. Act 1974: Enforcement Policy Statement. HSC(G)2, 17 July 1995.

22 Samuels A, 1994, Environmental Crime. *The Magistrate,* Vol. 50, No 2.

# Statutory notices 8

## 8.1 REGULATORY OPTIONS

There can be little doubt that one of the most useful tools available to the Environmental Health Officer is the statutory notice. A wide range of statutes provide for the service of legally enforceable notices under many differing circumstances. There are very sound reasons for this form of enforcement mechanism. The enforcement method of EHOs and others involved in environmental regulation has been typified by an informal regulatory approach. It is an approach which, as identified by Hutter (1988), relies on negotiation, bargaining, education and advice to secure compliance.[1] Such an approach is certainly not unique to Environmental Health Officers and has been found to apply to others involved in related areas of regulation (Hawkins (1984)[2]).

The Report of the Committee on Safety and Health at Work, the Robens Committee,[3] noted that in the area of health and safety at work the existing enforcement provisions involved the use of the criminal courts. The courts were, in the words of the committee, inevitably concerned more with events which have happened than with curing the underlying weaknesses which have caused them.

The Robens Committee recommended a more constructive means of ensuring that practical improvements are made and preventative measures adopted in the area of health and safety at work. Such measures should include the power to issue notices, the purpose here being the creation of 'Non-judicial administrative techniques for ensuring compliance with minimum standards of safety and health at work.' Though Robens was only concerned with safety in the workplace, its findings surely offer an insight into the themes underlying the availability and use of the statutory notice. The option a statutory notice offers is one of a number of points in a continuum of options available to an Environmental Health Officer to deal with unsatisfactory situations. It offers an intermediate step between inactivity and prosecution; it has the aim of resolving a problem without the need for resort to the courts and with the advantage of speed and cost savings. Despite its being an intermediate step it would be quite wrong to fail to appreciate the legal significance of the notice and the need to approach the selection, drafting and service of such documents with complete discipline and thoroughness. The conditions regulating the service of notices differ from statutory provision to statutory provision and it is not the purpose here to examine all such individual provisions in detail. It is incumbent on an Environmental Health Officer to ensure that he is familiar with the provisions he proposes to use. Here matters of general principle and concern will be examined, matters which are applicable to all notices.

Firstly it is important to remember that a notice can be served by an officer in his or her personal capacity, for example under s.10 of the Food Safety Act 1990 or s.20 of the Health and Safety at Work Act etc. 1974, or it can be served on behalf of the relevant authority, for example s.79 of the Environmental Protection Act 1990 or s.189 of the Housing Act 1985. In either case it will inevitably be the officer whose name is on the document who will, if there is an appeal, have to defend both the notice and his or her decision to serve it. One of the first points to note is that there is no such thing as an 'informal notice'. In the context used here a notice is a legally enforceable document served by an authorised body or person having legal effect, often capable of creating legally enforceable duties and backed up by legal sanctions. It is recognised that there exists a common practice of giving 'verbal notice' or of sending 'informal notices'. It should however be recognised that what is in fact happening is that the recipient of the information is merely being talked to or receiving a letter, which has no more legal significance than any other conversation or letter and most certainly is not a notice in the sense adopted here. It may be indicative of an ongoing state of affairs or of a course of conduct, but it is not a notice.

## 8.2 PRE-FORMAL ENFORCEMENT ACTION PROCEDURES

There has however recently been a significant change in the 'preliminaries' necessary before the service of certain notices. We have in the areas of health and safety at work[4], food safety and most recently housing[5] seen the introduction of the 'minded to' notice, with published studies suggesting that this move may be supported by those involved in enforcement.[6] This innovation clearly flows from the provisions of the Deregulation and Contracting Out Act 1994, which provides for the improvement of enforcement procedures and advocates codes of practice for regulatory agencies.

Typical of the approach is that now found in the Housing Grants, Construction and Regeneration Act 1996. Section 86 of the Act deals with the power of the Secretary of State to improve enforcement procedures. Under this section he may by order (The Housing (Fitness Enforcement Procedures) Order 1996) require that a local housing authority act in a specified way before taking action on any of the following kinds of enforcement action:

(a) serving a deferred action notice under section 81 or renewing such a notice under section 84;
(b) serving a notice under section 189 of the Housing Act 1985 (repair notice in respect of house unfit for human habitation);
(c) serving a notice under section 190 of that Act (repair notice in respect of house in state of disrepair but not unfit for human habitation);
(d) making a closing order under section 264 of that Act;
(e) making a demolition order under section 265 of that Act.

An order made under s.86 may provide that the authority:

(a) shall as soon as practicable give to the person against whom action is intended a written notice which
   (i) states the nature of the remedial action which in the authority's opinion should be taken, and explains why and within what period;
   (ii) explains the grounds on which it appears to the authority that action might be taken as mentioned in subsection (1); and
   (iii)states the nature of the action which could be taken and states whether there is a right to make representations before, or a right of appeal against, the taking of such action

and

(b) shall not take any action against him until after the end of such period beginning with the giving of the notice as may be determined by or under the order.

Accordingly the Housing (Fitness Enforcement Procedures) Order 1996 provides that before an authority takes any action against any person, they:

(1) shall give to that person a written notice stating –
   (a) that they are considering taking the action and the reasons why they are considering it; and
   (b) that the person may, within a period specified in the notice (not being less than 14 days), make either written representations to them or, if the person so requests, make oral representations to them in the presence of an officer appointed by the authority

(such a request being made not later than the expiry of 7 days beginning with the day on which that notice is given); and

(2) shall consider any representations which are duly made and not withdrawn.

For those involved in housing work the informal procedures are designed:

- to improve the transparency of the enforcement process;
- to help local authorities reach sensible decisions with owners; and
- to help reduce the burden that can arise from having to take formal enforcement action.

Additionally the order provides that the consequences of a failure to comply with its requirements is to create an additional ground for appeal over and above those specified in the relevant individual enforcement provisions.

This obligation, as is typical, does not preclude a local housing authority from taking immediate action in any case where it appears to them to be necessary to do so. Such an exemption exists in recognition of the fact that there may be instances where there is a need for immediate action and that such decisions can only properly be taken by an authority in the light of the circumstances of each case. While those circumstances will inevitably vary examples where immediate action might be required include:

- where an authority considers there is imminent risk to the health and safety of the occupants of the premises;
- where the management record of a landlord is considered unsatisfactory by an authority.

It would appear inevitable that authorities must now turn their minds to codes of practice which will inform their decisions in this area and enable them to be certain that they have regard not only to those things which they ought properly to consider when, for example, deciding the most satisfactory course of action for dealing with substandard housing, but also to whether or not they should be taking immediate action against any person (see Chapter 7). To simply adopt such guidance as is given centrally, without more thought or interpretation on its application in a particular authority, is not likely to be sufficient. For example guidance on the Housing Grants, Construction and Regeneration Act 1996 says it may be appropriate to take immediate action 'where an authority considers there is imminent risk to the health and safety of the occupants of the premises'. The definition of imminent risk is clearly open to interpretation and it will be up to an authority to decide what amounts to imminent risk, though in the context that if this wording is accepted it is the risk which must be imminent. The word 'imminent' here qualifies the word 'risk'. But the authority must decide if that then applies to any risk to health and safety however slight; should it be a risk of serious harm or something else?

### 8.3 WHO MAY RECEIVE A NOTICE

Each of the range of statutory notices available to the Environmental Health Officer requires to be served on a specified person or persons. A failure to correctly identify the recipient of a notice may render the notice void and frustrate the purpose of service. It is vital therefore that EHOs recognise correctly the recipients of such a notice.

#### 8.3.1 PUBLIC HEALTH ACT 1936 AND BUILDING ACT 1984

Under the Public Health Act 1936 and Building Act 1984 notices will commonly have to be served on the 'owner'. For example section 59 of the Building Act 1984 provides powers to a local authority to deal with defective drains etc. It allows a local authority to require, by notice, an owner of a building to make satisfactory provision for the drainage of that building. Owner is defined in s.343(1) of the Public Health Act 1936 (in terms the same as the Building Act) as 'the person for the time being receiving the rack rent of the premises in connection with which the word is used, whether on his own account or as agent or trustee for any other person, or would so receive the same if those premises were let at a rackrent'. Rack-rent is defined in s.126 of the Building Act as a rent that is not less than two-thirds of the rent at which the property might reasonably be expected to be let from year to year, free from all usual tenant's rates and taxes, and deducting from it the probable average annual cost of the repairs, insurance and other expenses (if any) necessary to maintain the property in a state to command such rent.

Because of the nature of drainage problems the question may arise here as to precisely which owners should receive notices. In *Swansea City Council v Jenkins and Others* (1994) it was held that where a private sewer was in such a condition as to be prejudicial to health or a nuisance under the provisions of the statute, the local authority was entitled to serve a notice to repair upon the owners of the property above the defect and not on all those who were served by the drainage system. Under section 59 of the Building Act discretion may be exercised to serve on the owner or occupier in certain circumstances.

### 8.3.2 HOUSING ACT 1985

Under sections 189 and 190 of the Housing Act 1985, dealing with unfit and substandard housing, the local housing authority are able to serve a repair notice (s.189(1)) on the person having control (defined in s.207) of the house. Where the premises consist of a flat which is unfit due to the condition of the building in which it is situated then the notice must be served on the person having control of the part of the building which requires the work (s.189(1A)). Where the house is in multiple occupancy the notice can be served, in the alternative, on the person managing the house (s.189(1B)). A copy of any repair notice must be served on others having an interest in the premises whether as freeholder, mortgagee or lessee (s.189(3)). Similar provisions exist to deal with housing which though fit suffers from substantial disrepair (s.190). The 'person having control of the house' is the person who receives the rack-rent of a house, whether on his own account or as agent or trustee for any other person, or who would so receive it if the house were let at a rack-rent. 'Rack-rent' means rent which is no less than two-thirds of the full net annual value of the premises (s.207 Housing Act 1985). Where a local authority serves a notice in respect of part only of a building then the local authority must serve notice on the person who in their opinion ought to execute the works contained in the notice.

Section 352 of the Housing Act 1985 empowers a local authority by the service of a notice (s.352(2)) to take action to ensure that a house is fit for multiple occupation. Such a notice may, as for the repairs notices, be served on the person having control of the house (s.398 Housing Act 1985) or on the person managing the house (see also s.79 Housing Act 1996). When served the local authority must inform any person who is to their knowledge the owner, lessee, occupier or mortgagee, that the notice has been served (s.352(s)). Section 354 of the Housing Act 1985 deals with the power to limit the number of occupants of an HMO. Not less than seven days before giving a direction an authority must serve a notice of intention on:

- the owner of the house
- every person known to be a lessee of the house, and
- post the notice in some position in the house where it is accessible to those living in the house.

An authority must, within seven days from the giving of the direction, serve a copy of the direction on:

- the owner of the house
- every person known to be a lessee of the house, and

- post a copy of the Direction in some position in the house where it is accessible to those living in the house.

Similar requirements are to be found in relation to s.358 of the Housing Act 1985 dealing with overcrowding.

Of course on no occasion can a local authority serve notice on itself as landlord. In *R v Cardiff City Council ex parte Cross* (1983) it was held by the Court of Appeal that a local authority cannot take action to deal with unfit houses or houses in substantial disrepair where that house is owned and managed by that authority.

### 8.3.3 HOUSING GRANTS, CONSTRUCTION AND REGENERATION ACT 1996

Prior to the Housing Grants, Construction and Regeneration Act 1996, if satisfied that a house was unfit a local authority was obliged to undertake one of four courses of action.

- Clearance area (s.289 Housing Act 1985), or a
- Demolition order (s.265 Housing Act 1985), or a
- Closing order (s.264 Housing Act 1985), or a
- Repairs notice (s.189 Housing Act 1985).

Now in Chapter IV Part I of the 1996 Act there is a fifth option: the Deferred Action Notice. Section 82 of the Act deals with the service of a deferred action notice and provides that for a dwelling-house the notice is served on the person having control of the dwelling-house in accordance with s. 207 of the Housing Act 1985.

For a house in multiple occupation the deferred action notice is served on the person having control of the house as defined in s.398 of the Housing Act 1985; this is the person who receives the rack-rent of the premises, whether on his own account or as agent or trustee for another person, or who would receive it if the house were let at such a rack-rent. Section 82(3) does allow the authority to serve the notice on the person managing the house in multiple occupation instead of the person having control of the house.

Where a dwelling which is a flat or a flat in multiple occupation is unfit by virtue of section 604(2) of the Housing Act 1985 (the addition of subsection 2 attributable to the Local Government and Housing Act 1989), the notice must also be served on the person having control of the building or part of the building in question.

Where an authority serves a deferred action notice it must serve a copy of the notice on any other person having an interest in the dwelling-house or house concerned, whether as freeholder, mortgagee or lessee, and they may serve a copy on anyone with a licence to occupy the premises (s.82(4)).

### 8.3.3.1 The Deferred Action Notice

The Deferred Action Notice is, at the moment, unique amongst the notices which may be served by an Environmental Health Officer, in that it does not actually require the recipient to do or refrain from doing anything. Section 81(1) of the Housing Grants, Construction and Regeneration Act 1996 stipulates that if the local housing authority are satisfied that a dwelling-house or house in multiple occupation is unfit for human habitation, but are satisfied that serving a deferred action notice is the most satisfactory course of action, they must serve such a notice.

Guidance on the deferred action notice option states that it has been introduced:

- to assist authorities in the exercise of their fitness enforcement duties having regard to the discretionary renovation grant system introduced under the 1996 Act;
- to provide authorities with additional flexibility to develop and implement strategies for their finite resources;
- to enable authorities to respond more readily to the wishes of those who might not want to face the upheaval that making their homes fit might entail in cases where those wishes, when weighed with all the other relevant factors, point to deferred action as being the most satisfactory course of action.

The Secretary of State in providing guidance states an expectation that local authorities will use the deferred action notice option sensibly and in relation to the degree of formal enforcement activity an individual local authority is already undertaking.

Examples given of where the service of a deferred action notice might be appropriate include:

- where unfit premises are in an area designated by an authority for renewal in the future;
- where an authority wishes to respond affirmatively to an application for a renovation grant to make the premises fit but does not have sufficient funds in the current financial year;
- where the elderly home owner might welcome minor works of improvement but not the upheaval that making the premises fit might entail;
- where the nature of unfitness is not considered by the authority to be seriously detrimental to the wellbeing of the occupants.

By s.81(2) the Deferred Action Notice must:

(a) state that the premises are unfit for human habitation;

(b) specify the works which, in the opinion of the authority, are required to make the premises fit for human habitation; and

(c) state the other courses of action which are available to the authority if the premises remain unfit for human habitation.

The Housing (Deferred Action and Charge for Enforcement Action) (Forms) Regulations 1996 prescribe the forms for use by local housing authorities in this regard.

A notice becomes operative, if no appeal is brought to the County Court (s.83), on the expiry of 21 days from the date of the service of the notice and is final and conclusive as to matters which could have been raised at appeal.

The serving of a deferred action notice though not actually requiring any works to be done does not prevent a local authority from taking any other course of action in relation the premises at any time.

Section 84 of the 1996 Act deals with the review of Deferred Action Notices. A local authority may review a deferred action notice at any time, and must do so not later than two years after the notice becomes operative and at intervals of not more than two years thereafter. The review must include a reinspection of the premises and if the most satisfactory course of action in respect of the premises remains a Deferred Action Notice then the authority must renew the notice. It remains to be seen if the idea of Deferred Action Notices will surface in any other area.

### 8.3.4 ENVIRONMENTAL PROTECTION ACT 1990

Under s.80 of the Environmental Protection Act 1990, where a local authority is satisfied that a statutory nuisance exists or is likely to occur or recur in its area it must serve an abatement notice on the 'person responsible' requiring the abatement of the nuisance or prohibiting or restricting its occurrence or recurrence. It can also require the execution of works or other steps necessary for the purpose of the notice. By section 79(7)(a) 'person responsible' means the person to whose act, default or sufferance the nuisance is attributable. Where the nuisance arises from any defect of a structural character the abatement notice must be served on the owner of the premises (s.80(2)(b)). Where the person responsible cannot be found or when the nuisance has not yet occurred the notice must be served on the owner or occupier of the premises (s.80(2)(c)).

### 8.3.5 FOOD SAFETY ACT 1990

Section 10 of the Food Safety Act 1990 provides that if an authorised officer of an enforcement authority has reasonable grounds for believing that a proprietor of a food business is failing to comply with food hygiene or food processing regulations, he may serve a notice on the proprietor requiring him to take measures to remedy the defects within a given period. The power here is limited to service on the proprietor. Proprietor is defined in s.53 of the Act as the person by whom the business is carried on. This section confers a discretionary power on officers to serve an Improvement Notice to secure compliance with food safety and hygiene regulations, which are specified in s.10(3).

### 8.3.6 HEALTH AND SAFETY AT WORK ACT 1974

Under section 21 of the Act, if an inspector is of the opinion that a person is contravening one or more of the relevant statutory provisions or that there has been a breach in the past and circumstances make it likely that the contravention will continue or be repeated then he may serve on that person an Improvement Notice. A notice under s.21 is served on the person responsible for the breach of the legal requirements. A prohibition notice under s.22 of the Act is served on the person carrying on or in control of the activities concerned, whether or not that person would also be responsible for any breach of legal requirement. There is clearly here, unlike most of the other provisions, no requirement for ownership of or interest in land etc.

## 8.4 SERVICE ON PERSONS OTHER THAN INDIVIDUALS

There will be many occasions when an officer will find himself dealing, not with an individual, but with a company or other such organisation. As a matter of law this need cause little concern as a body recognised as having a legal personality can certainly be a person capable of receiving a statutory notice. Notices are often to be served on 'Persons' and the Interpretation Act 1978, s.5 and Schedule 1, defines a 'Person' as including a body or persons corporate or unincorporate. This is to be contrasted with the term 'individual' which statutory draftsmen will tend to use when not wishing to include such organisations. The problem which often confronts the Environmental Health Officer is not the legal one of whether they can serve notice on a body corporate or unincorporate, but the practical one of recognising what the organisation before them is; is it corporate or unincorporate? There is more than one type of corporation. The most common may be defined as the legal entity formed by registration under the Companies Act 1985 and its predecessors. This is often called a 'company' though more correctly it is a registered company. Its name will end in the word 'Limited' (Ltd) or 'plc' (public limited company). It is important to recognise this distinction as there may be many other organisations which will use the word 'company' in their name but remain unincorporated, e.g. partnerships and sometimes even sole traders. Registration under the Companies Act is not the only means by which a corporation may be created. It is also possible to achieve incorporation by a Royal Charter granted by the Crown. This is, for example, the way that universities are incorporated. Corporations, however formed, are recognised as having a distinct legal personality. They exist independently in law. They can own property, enter into contracts and acquire liability, quite independent of their membership. They are an independent, though artificial, 'legal person'. In the words of Lord Macnaughton in *Salomon v LA Salomon & Co. Ltd* [1897]:

> When the memorandum is duly signed and registered, though there are only seven shares taken, the subscribers are a body corporate capable forthwith ... of exercising all the functions of an incorporated company ... that company attains maturity at birth. There is no period of minority – no interval of incapacity ... The company is at law a different person altogether from the subscribers to the memorandum.

In *H L Bolton (Engineering) Co. Ltd* v *P J Graham and Sons Ltd* [1957] Lord Denning likened a corporation to a human body: having a brain, a directing mind, which gives orders and limbs which carry out those orders. The 'directing mind' being the board of directors and senior managers, they control what the company does. The limbs are the employees and other agents who carry out the orders. When an Environmental Health Officer finds himself dealing with such organisations, then notices should be served on the body by name. This can be achieved by serving notice on the secretary to the company (s.233(3) Local Government Act 1972) at the registered office of the company, if that office is in the UK. If there is no registered office in the UK then service may be at any place in the UK where the corporation trades or conducts its business.

For unincorporated bodies other considerations apply. For a partnership, defined in s.1 of the Partnership Act 1890 as 'the relationship which subsists between persons carrying on a business in common with a view of profit', the situation is different. A partnership does not have a separate legal personality as does a corporation. Therefore a notice should be served on each partner. Although each partner in a partnership has liability it is nevertheless good practice to serve all with a notice; there can then be no later suggestion that a failure to comply with a notice was due to their never having received it. It may be appropriate to record all recipients on the face of each such notice served; this may result in quite a long list of names appearing on each notice served but will have the advantage of making it clear to all recipients that identical notices have been served on all partners. Section 233(4) of the Local Government Act 1972 stipulates that for a partnership a notice should be served on a partner, or a person having the control or management of the partnership business, at the principal office of the partnership.

For other unincorporated associations e.g. clubs and societies, at common law such an organisation has no legal identity and cannot acquire criminal liability. However as the Interpretation Act now defines a 'person' as including a body of persons corporate or unincorporate, then, since 1979, it would appear that where a criminal matter refers to a person it will also refer to an unincorporated body. Indeed there are some statutory offences which expressly provide that they may be committed by unincorporated associations (see Trade Union and Labour Relations Act 1974). There would therefore appear to be no problem with serving a notice on an unincorporated association since it is a 'person' and can receive such a notice. The problem in fact will arise later with the enforcement of that notice via criminal procedures. Enforcement often relies on being able to bring the recipient of a notice before the courts for non-compliance with the terms of the notice. To facilitate this the only sensible solution would be to serve a notice on all members of the body, who could then be required to appear before a court and answer any charges relating to non-compliance with a notice. For partnerships these members are, as stated, all partners. For other associations this may mean all members. Service on all members of an unincorporated body may often not be a practical proposition. An EHO contemplating serving notice on all members of a golf club, for example, should think very carefully before undertaking such a procedure. It carries with it the very real possibility of appearing to be excessively bureaucratic and would be practically unenforceable in any event. Instead of such an approach the officer would be advised to consider service on all who could be so closely identified with the offending activity that they could be deemed to be in control of and therefore responsible for it. This may mean that it is all the members of a particular committee who should receive the notices. Identifying such a committee may not be an easy task for an Environmental Health Officer and may require him to first familiarise himself with the constitution of the body in question. Though this may not be an easy matter, and will certainly require additional work, there is nevertheless little viable alternative. It is likely to be as successful, in terms of outcome, as service on all members and result in less adverse publicity than the unrealistic service of several hundred notices on the membership of an entire club.

## 8.5 NOTICES AND COPIES

It should be borne in mind that there is only ever one notice and that is served on the person concerned. All the other documents, whether held on file, posted up, or sent for information are copies. It goes without saying that it is vital that all such copies are true and accurate copies of any original sent.

## 8.6 IDENTIFYING THE RECIPIENT

Self-evidently one of the most basic matters to have regard to is ensuring that any notice served goes to the correct person, that is the person legally capable of receiving such a notice. Yet, even when it is known to which class of persons a notice should be sent, ensuring such a person is the actual recipient may still prove to be a problem. In the case of *Courtney-Southan v Crawley UDC* [1967], Mr and Mrs Courtney-Southan ran the business of 'car parkers' on Tinslow farm, near to Gatwick airport. The land used was owned by Mrs Courtney-Southan. Though the business was run jointly by both parties, Mr Courtney-Southan, by agreement, exercised control over the business and received all moneys which he paid into his account. This was ultimately divided between himself and his wife.

On 30 July 1963 Mr Courtney-Southan applied for planning permission to use the land for the storage of cars. This application was refused. A second application was made in October, which was also refused. Consequently Mr Courtney-Southan appealed. On each application and appeal Mr Courtney-Southan had certified either directly, or via his solicitor, that he was the owner of the land. A similar statement was also made by him to a local enquiry. Despite the appeal permission was still refused. Mr Courtney-Southan continued to use the land for the same purpose and was ultimately served with an enforcement notice in May 1965. This notice was addressed to him by name as the owner and required that he discontinue using the land to store cars. The notice was not complied with and therefore a prosecution was commenced for non-compliance with the notice. At that hearing Mr Courtney-Southan submitted that not he, but his wife, was in fact the owner and that as the enforcement notice had not been served on her it was of no effect.

Essentially he was not a person entitled to receive the notice and it was therefore of no effect. He was, despite this, convicted of the offence but appealed. On appeal it was held that Mr Courtney-Southan was not the owner of the land and therefore the service of notice did not comply with requirements of the Act in question. His conviction for breaching the terms of the enforcement notice was quashed. From a local authority perspective this can appear an unfair decision. The court was aware of this, Widgery J noting (at 248) that: 'In view of the history of the matter and the appellant's repeated assertion of ownership it is not surprising that the respondents treated him as the sole owner and occupier.'

Nevertheless it was a condition precedent to the validity of the notice that there be proper service of the notice on the owner and occupier – this had not been met and the earlier conviction had to be quashed. It is tempting here to ask what more the authority could have done. It had on numerous occasions and under 'formal' circumstance been assured that Mr Courtney-Southan was the owner. Though this is true it is also irrelevant. The case is not about blame, it is about proof. No one here could prove that a valid notice had been served on a person entitled to receive it and on this the case was decided.

The question then arises: what can an Environmental Health Officer faced with serving notices do to ensure those notices go to the correct individual? Section 16 of the Local Government (Miscellaneous Provisions) Act 1976 offers one answer. This section deals with power of local authorities to obtain particulars of a person's interest in land and can prove invaluable in identifying who should properly receive a notice. Section 16 states that where, with a view to performing a function conferred on a local authority by any enactment, the authority considers that it ought to have information connected with any land, then the authority may serve a notice on the occupier of the land, or on any person who has an interest in the land either as freeholder, mortgagee or lessee or who directly or indirectly receives rent for the land, or on any person who, in pursuance of an agreement between himself and a person interested in the land, is authorised to manage the land or to arrange for the letting of it. Such a notice must specify the land, the function to be performed by the local authority and the enactment which confers the function. The notice then acts so as to require the recipient of the notice to furnish to the authority, within a period not less than 14 days, information on the nature of his interest in the land. The recipient must also provide the name and address of each person who he believes is the occupier of the land and of each person who he believes is, as respects the land, any person with an interest in the land or any person authorised to manage the land or arrange for its letting.

Typically such a notice will seek information on a person's interest in the land, the name and address of the freeholder and the address of any occupier, whether that occupier holds on a lease, and the name and address of person who receives the rent. It may also seek information on the mortgagee. In short it permits an Environmental Health Officer to gain essential information preparatory to the service of any notice. It is worth noting that definition of the word 'functions' for the purpose of s.16 includes powers and duties (s.44(1)) which gives this section a wide applicability and makes it an extremely useful tool. It is a means of obtaining information backed by sanction for those who do not comply in one of two ways. Section 16(2) states that any person who fails to comply with the requirements of such a notice or in furnishing any information in compliance with such a notice makes a statement which he knows to be false in a material particular or recklessly makes a statement which is false in a material particular shall be guilty of an offence and liable to a fine not exceeding level 5 on the standard scale (currently £5000).

Though very valuable, the effectiveness of the s.16 notice can be limited by the way in which such notices are managed. It is for example standard practice to include a return pro-forma for the recipient of such a notice to complete and return to the relevant local authority. Unfortunately all too often these reply sheets are written in a form of English almost incomprehensible to the ordinary public. The result is that forms are returned incomplete or inaccurately completed and do not then achieve the purpose for which they were designed. This problem can be reduced by more careful design of the s.16 response document, an example of which is reproduced. With appropriate adaptation this could assist the respondent in understanding better what is required of them and the relevant authority in obtaining a better quality of information.

### 8.7 HER MAJESTY'S LAND REGISTRY

As long ago as 1897 the Land Transfer Act of that year allowed county councils to require compulsory registration of land in their area on the sale of that land. Later the Land Registration Act 1925 gave the Privy Council power to declare new areas of compulsory registration without the need for a request from a county council. That Act however stipulated that this power was not to be not be exercised for 10 years. The provision became effective therefore only in 1936. In the years since then compulsory registration has been gradually extended so that there are now approximately 14.7 million registered properties in England and Wales. Though it is estimated that there remain approximately 7.3 million unregistered properties. The registration of title is carried out by Her Majesty's Land Registry which is made up of a headquarters at Lincoln's Inn Fields in London and 19 district registries across the country.

The Registry deals with:

- First Registrations – this is where unregistered land is brought into the registration system;
- Transfers of Whole – this is where all of the land comprised in a title is transferred to another party or parties;
- Transfers of Part where a subdivision of land occurs. This would occur for example when land is divided into building plots;
- Other dealings with registered land, e.g. mortgages.

**Property Reference ...........**
**LOCAL GOVERNMENT (MISCELLANEOUS PROVISIONS) ACT 1976**

**Please return to:-**

**This form must be returned within ..... days from the date of service upon you**

In reply to your Notice dated the ... day of ........... 199.. requiring me to give you certain information as to my interest and the interest of others in the property known as:

.....................................................................................................................................

I HEREBY STATE that the answers to the questions set out in the Schedule below comprise a true and correct statement of all the information required by that Notice, so far as that is within my knowledge.
Dated this.....day of .............199...

|  | **First Name (s)** | **Surname** |

**FULL** Name (BLOCK CAPITALS) ........................................................................
**Address** (BLOCK CAPITALS)

.....................................................................................................................................

**Signed**........................................................

**Schedule**

| Question | Answer (please write in block capitals) Do not leave any boxes blank; if they do not apply write 'none'. |
|---|---|
| 1. What is the nature of your interest in the aforementioned property? e.g. Freeholder, leaseholder, tenant, mortgagee | |
| 2. Give the *FULL* name and address of each person who you believe is the occupier of the said property **(give full names and not simply initials)** | |
| 3. Give the *FULL* name and address of any other person who has an interest in the said property either as **a)** Freeholder (Owner) **b)** Mortgagee (Bank, Building Society etc.) **c)** Lessee (Tenant) or **d)** Who directly or indirectly receives rent for the said property (Agents etc.) | **a)** **b)** **c)** **d)** |
| 4. Give the *FULL* name and address of any person who in pursuance of an agreement between himself and a person interested in the said property is authorised to manage or to arrange for the letting of it | |
| 5. Give the *FULL* name and address of any other person having an interest in the premises. State the nature of such interest | |

**Form 8.1** Local Government (Miscellaneous Provisions) Act 1976

Since December 1990 the register of title has been a public register capable of being inspected by anyone. When attempting to gain ownership details of a property access to this register may prove invaluable, as registered title is taken to be an official record of the up-to-date ownership of the property. Also since it discloses some details regarding mortgages and other charges on the property this can be most useful in establishing who might be an appropriate person to contact when considering the service of a notice under s.16 of the Local Government (Miscellaneous Provisions) Act 1976.

### 8.7.1 THE FORM OF THE REGISTER

The register for an individual title is, in accordance with Rule 2 of the Land Registration Rules 1925, divided into three parts with an associated plan.

### 8.7.1.1 The Property Register

The content of this part is governed by Rule 3 and is required to contain a description of the land and the estate comprised in the title and to state any rights which benefit the land. The Land Registration Act 1925 and the Land Registration Rules 1925 regulate the activity of the Land Registry with regard to description of property given.

Rule 272 of the Land Registration Rules 1925 requires that 'the Ordnance Map shall be the basis of all registered descriptions of land' and Rule 273 provides that:

(1) For the purpose of describing land, there shall be prepared and kept in the Registry a series of maps which together shall be called the Land Registry General Map.
(2) Each of the series shall be either:
    (a) an extract from the Ordnance map revised and corrected to such extent as may be necessary; or
    (b) a map based on and uniform with the Ordnance Map and so constructed that every parcel shown on it can be accurately located on the Ordnance Map.

This permits the Registry to identify any piece of registered land by reference to a filed plan on which is indicated (usually by red edging) the extent of the land in the title. The land, under a given title number, is identified in the property register by the use of the words: 'The Freehold land shown and edged with red on the plan of the above title filed at the Registry and being known as ... '. In the registered conveyancing system, it is this filed plan which constitutes the official title plan. There may also be in this part of the register information regarding any rights benefiting the land e.g. easements.

### 8.7.1.2 The Proprietorship Register

This part is governed by Rule 6 of the Land Registration Rules 1925 and states the class of the title, the name of the proprietor and the details of any caution, restriction or inhibition affecting his power of disposition, i.e. to sell, mortgage etc. the property.

### 8.7.1.3 The Charges Register

This part, regulated by Rule 7 of the Land Registration Rules 1925, contains registered mortgages and notices of other encumbrances which adversely affect the title. Here may be found details of restrictive covenants, rights of way etc.

### 8.7.1.4 Land and Charge Certificates

A copy of the entries on the register and of the filed plan is given to each registered proprietor as a land certificate. Where there is a charge this land certificate is kept in the Registry and the mortgagee is issued with a charge certificate. The mortgagor will then hold nothing and cannot therefore do anything without the mortgagee's consent.

A most useful publication in this area is 'Explanatory Leaflet Number 15: The Open Register – A Guide to Information Held by the Land Registry and How to Obtain It'. This gives details on registered title, the extent of the information held, how it may be inspected and the fees that are charged, the locations of all District Registries and the counties for which they are responsible; there is also reproduced a specimen register. If all that is required is the name and address of the registered owner of freehold or leasehold land identified by a single postal address then this booklet contains Land Registry Form 313 which may be used to request this information. Copies of this booklet are available from any district land registry or from the Land Registry at 32 Lincoln's Inn Fields, London, WC2A 3PH.

### 8.8 COMPANIES

For notice to be sent to companies there are a number of options for finding out where to serve a legal notice. There are published references available in commercial libraries, for example Dunn and Bradstreet publish *Key British Enterprises* which gives the full legal name and headquarters address of the top 50 000 British companies. For corporate bodies there are also the documents filed with the Registrar of Companies. The Companies Registry for England and Wales, which is an executive agency in the Department of Trade and Industry, is to be found at Companies House, Crown Way, Maindy, Cardiff CF4 3UZ. There are also main offices in London and Edinburgh and satellite offices in Birmingham, Glasgow, Leeds, and Manchester. Here it is possible for the public to examine copies (see s.709 Companies Act 1985) of the files of the Registrar relating to more than 1 000 000 companies. Such files will contain amongst other information the following:

- The Memorandum of Association
- The Articles of Association
- The address of the registered office
- Details of the names and addresses of directors and company secretary.

The main offices are able to provide company searches on a same day basis, with the satellite offices providing a next day service. It is possible to search by post, fax or telephone, though it is also possible to use one of the Search Rooms in the main and satellite offices. Here there exists the facility to carry out enquiries via a computer search facility. This is able to provide information on, amongst other things, the registered office address of the company, the type of company, the list of directors and date of incorporation. There is a scale of fees for all the services offered by Companies Houses, which are modest, and all clearly given in the literature produced by the agency.

Also of use when trying to obtain information on a company are the provisions of the Business Names Act 1985. By s.1 of the Act certain rules with regard to disclosure apply to persons carrying on a business in Great Britain under a name which does not consist of:

(a) In the case of an individual, his surname;
(b) In the case of a partnership, the surnames of all partners who are individuals and the corporate name of all partners who are corporate bodies;
(c) In the case of a company, its corporate name.

Section 4(1) of the Act stipulates that for any of the above named persons they must:

(a) State legibly on all business letters, order forms, invoices, demands for payments and receipts:
    (i) In the case of a partnership, the name of each partner;
    (ii) In the case of an individual, his name;

(iii)In the case of a company its corporate name; and

(iv)For each an address in Great Britain at which service of documents relating to the business will be effective;

(b) In any premises to which suppliers or customers have access, display prominently a notice containing the names and addresses referred to in (a) in a place where it may easily be read; and

(c) Supply in writing the names and addresses referred to in (a), to any person with whom anything is done or discussed in the course of business who asks for such detail.

Section 4(3) of the Act gives exceptions to the above in the case of partnerships of more than 20 persons. A failure to comply with the provisions of the Act is a criminal offence.

## 8.9 SERVICE WHERE THE RECIPIENT CANNOT BE FOUND

Despite the provisions of s.16 and the availability of the land and company registries there will be occasions where, despite his or her best efforts, an officer will not be able to trace the relevant person who should receive the notice. In such cases where the name or address of any owner, lessee or occupier of land cannot after reasonable enquiry be ascertained, then under the provisions of s.233(7) of the Local Government Act 1972 a notice may be served either by leaving it in the hands of a person who is or appears to be resident or employed on the land, or by leaving it conspicuously affixed to some building or object on the land. It should not, when relying on this section, simply be put through the letter box of an empty building; it must be 'left in the hands' of someone or be 'conspicuously affixed'. This should always be viewed as the least favoured option but one which nevertheless may be satisfactorily employed. If the notice is given to someone on the property details of that person should be obtained and recorded as for normal personal service of notices. If the notice is affixed to the property it is common practice for photographs to be taken showing the notice and its position on the premises. Officers should also contemporaneously record details regarding day, date, time and location of any notice so served.

## 8.10 FORMAT OF A NOTICE

For many notices the enabling statutory provision will prescribe the content of the notice. Where such is the case then that format must be followed. For example an improvement notice under s.21 Health and Safety at Work Act 1974 must record:

(a) that the officer is of the opinion that a person is contravening one or more of the relevant statutory provisions or that there has been a breach in the past and circumstances make it likely that the contravention will continue or be repeated
(b) the relevant statutory provision contravened
(c) his reasons for believing that there is or has been a breach
(d) a requirement to remedy the contravention
(e) the time period within which there must be compliance with the notice.

Similarly s.10 of the Food Safety Act 1990 requires that an improvement notice, under that section, should:

(a) state the officer's grounds for believing that the proprietor is failing to comply with the relevant regulations
(b) specify the matters which constitute failure to comply
(c) specify the measures which the proprietor must take to secure compliance
(d) require the proprietor to take those measures, or measures which are at least equivalent, within such period, being not less than 14 days, as may be specified in the notice.

The requirement may go beyond merely prescribing the content of a notice and extend to prescribing the form. Thus for a notice under s.10 of the Food Safety Act 1990, the Food Safety (Improvement and Prohibition Prescribed Forms) Regulations 1991 provide a format that such a notice must take. In similar manner the Housing (Prescribed Forms) (No 2) Regulations 1990 provide prescribed forms for a number of the notices which may be served under the Housing Act 1985. It is obligatory on an officer to ensure that he uses the correct form in all appropriate circumstances. Where the form is prescribed it must be used but not all notices have prescribed forms. There is not for example a prescribed form for an abatement notice under s.80 of the Environmental Protection Act 1990, nor does there exist provision for the making of any. Where an officer is in doubt as to the format of any proposed notice he would be well advised to consult one of the Encyclopaedias on Forms and Precedents or many of the pre-prepared forms published by Shaw & Sons Ltd.

8.10.1 ENGLISH

The use of an acceptable standard of English is an obvious requirement of any statutory notice. Yet often out of an understandable and entirely correct desire to be precise the wording of a notice can become too complex and so fail to be good English. In *Myatt v Teignbridge DC* [1995] the notice in question was an abatement notice under s.80 of the

Environmental Protection Act 1990 dealing with noise from dogs. The notice said:

> Take notice that under the provisions of the Environmental Protection Act 1990 the Teignbridge District Council being satisfied of the likely recurrence of a statutory nuisance under section 79(1) of that Act at the premises known as 4 Rock Cottages, Milbury Lane within the district of the said council arising from the keeping of dogs.
>
> Require you, as the person responsible for the nuisance, within 28 days from the service of this notice
>
> Hereby prohibit the recurrence of the same and for that purpose require you to cease the keeping of dogs ...

Lord Justice Butler-Sloss felt that even though it was not the point before the court, the notice did not show a very good use of English and could certainly be improved. Indeed the local authority accepted that the wording was unfortunate. Though in this case the court identified a lack of care in drawing up the notice this was not held fatal to its validity; nevertheless one can empathise with the discomfort which may have been felt when this deficiency was first identified. Increasingly the work of the Environmental Health Officer is coming under greater scrutiny. The Environment and Safety Information Act 1988, which has been effective since the 1 April 1989, requires a local authority to maintain a register open to the public of all notices served. Though this does not extend to the Public Health or Housing Acts it does include Improvement and Prohibition notices under the Health and Safety at Work etc. Act 1974. If notices are to be held on public registers it is as well that they give an impression of professionalism in their use of language.

### 8.10.2 COMPLIANCE WITH RELEVANT STATUTE

Once the problems of identifying the correct recipient and the correct prescribed form have been overcome it may still be the case that relatively minor errors in draftsmanship may still have surprisingly severe consequences. Competent draftsmanship based on an understanding of the relevant legal provisions is vital to the proper creation and, if necessary, enforcement of any statutory notice. In *Perry v Garner* [1953] Perry proffered an information charging Garner, under s.5(2) of the Prevention of Damage by Pests Act 1949, with failing to take the steps required by a notice served by Wycombe Rural District Council under s.4(1) of the Act. On hearing the case the justices initially dismissed the information.

On appeal Lord Goddard noted that s.4 of the Prevention of Damage by Pests Act 1949 stated that if, in the case of any land, it appears to the local authority that steps should be taken for the destruction of rats and mice on land or for otherwise keeping land free from rats and mice they may serve on the owner or occupier of the land a notice requiring him to take, within a reasonable period as may be specified in the notice, such reasonable steps for the purpose aforesaid as may be specified. He went on to state that: 'The notice in this case required the owner to take the following steps "poison treatment of infested land, such measures to be carried out to the approval of the local authority, or other work, of a not less effectual character, to be executed for the destruction of the rats." '

Lord Goddard found this not to be specifying the steps which are to be taken. The notice was found to specify a step which the owner may take and then tell him he may take other steps which were not specified. The thing required to be done at once then became unspecified. The notice was found to say to the owner to do a particular thing or something else and the something else was left completely at large. The court therefore held that the notice did not comply with the requirements of the section of the Act. If it had confined itself to poison treatment, it would have complied with the section; since it did not confine itself to poison treatment, but said that the owner is to put down poison or do something else, which it did not specify, it was not a good notice under the Act.

An Improvement Notice issued under s.21 of the Health and Safety at Work etc. Act 1974 should identify the relevant statutory provision which is being contravened by making reference to the particular part of the legislation i.e. the section, subsection and paragraph. If this is not identifiable then the notice must include particulars of the contravention. In *West Bromwich Building Society v Townsend* [1983] an inspector, in a notice, gave his opinion that there was a breach of s.2(1) of the Health and Safety at Work etc. Act 1974 (the general duty of employers to employees) without giving particulars of the breach. This led to the notice being quashed.

The Divisional Court, in *London Borough of Bexley v Gardner Merchant Ltd* [1993], held that the provisions of s.10 of the Food Safety Act 1990 should be seen as stipulating that a notice has four separate requirements. These are to be found in s.10(1)(a), s.10(1)(b), s.10(1)(c), and s.10(1)(d). A failure to comply with part of s.10 would not, in the opinion of the court, be cured by compliance with other parts of it. When a notice of this sort is not in accordance with the statute it must not be amended or modified but must be cancelled. As has already been stated s.10 of the Food Safety Act 1990 requires that an improvement notice should: 'State the officer's grounds for believing that the proprietor is failing to comply with the relevant regulations and specify the matters which constitute failure to comply.'

To assist in the correct construction of such notices Food Safety Act 1990, Code of Practice No. 5, stipulates that the wording of the notice should be clear and easily understood. The improvement notice must include details of the regulation contravened and the reason for the opinion of the authorised officer that there has been a contravention. It is not sufficient simply to quote the regulations. In the instant case the second paragraph of an Improvement Notice under s.10 (dealing with the officer's grounds for believing there was a contravention) stated that the proprietor failed to comply with regulation 18 of the Food Hygiene (General) Regulations 1970 (see now Sch. 1, Ch. 1, paras 3 and 4 of the Food Safety (General Food Hygiene) Regulations 1995) because: 'a conveniently accessible wash hand basin with a supply of hot and cold water or hot water at a suitably controlled temperature was not provided to the serving area.'

Examination of this paragraph of the notice reveals that it does not specify the matters which constitute the failure. For example on this wording the failure could be:

(a)  The wash hand basin is absent
(b)  The wash hand basin is present but not accessible to the serving area
(c)  The wash hand basin is present, it is accessible, but not conveniently so, to the serving area
(d)  The wash hand basin does not have a supply of hot water
(e)  The wash hand basin does not have a supply of cold water
(f )  The wash hand basin does have a supply of hot water but it is not at a suitably controlled temperature.

There are of course other possible combinations of the above but the point is made.

When drafting a notice the requirement to specify means just that, to be specific. An insight into the approach adopted by the courts can be found in the case of *Canterbury City Council v Bern* (1981). The respondent, Mr Bern, was the occupying tenant of a building which lacked standard amenities. He wrote to the local authority asking them to initiate action to improve his dwelling under section 90 of the Housing Act 1974. (See now s.189 Housing Act 1985.) Subsequently the local authority notified the owner of the property and served a notice on him under s.90 of the 1974 Act. The owner put a wash-basin and WC in the property and asked the council to put in a bath. However, Mr Bern objected when it was proposed that the bath should be placed in a bedroom. He subsequently prevented the council's workmen from getting into the house and was prosecuted for obstruction. Whether or not he was guilty of this offence depended on whether the notice was valid. If it was not the council had no power to enter and he was therefore within his rights to obstruct them. The Divisional Court found that once again the notice did not 'specify' the works. Merely to require a person to provide a list of amenities is to fail to specify any works at all.

Forbes J said that there were two points of law at issue:

1. Whether the notice complied with the statutory requirements for such notices
2. Whether, if it did not, its validity was open to challenge at the present stage.

The operative part of the notice was:

> The council ... hereby requires you to carry out within 12 months ... the works specified in the schedule to this notice ... Provide a fixed bath or shower in a bathroom. Provide a wash-hand basin. Provide a hot and cold water supply to fixed bath or shower. Provide a hot and cold water supply to wash-hand basin. Provide a hot water supply to a sink. Provide a water closet accessible from within the dwelling.

The question before the court was whether, in that form, the notice specified 'the works ... required to improve the dwelling to the full standard' within section 90(1). Forbes J held that it did not. It was found that all that the council had done was to add the imperative of the verb 'provide' in front of each item in the list of amenities in Schedule 6 to the 1974 Act. In the words of the judge:

> By no stretch of the imagination could it be said to be a specification of works. If Parliament had meant an improvement notice to specify merely which of a list of standard amenities that the person in control of the dwelling had to provide it would have said so. It did not, but chose to say that the council had to specify works. Those were two wholly different requirements, and the notice, because it only required the recipient to provide a list of amenities, failed to specify any works at all. For those reasons the document served on the present case was not an improvement notice under the Act.

The court went on to note that from the statutory provisions, and from the decision in *West Ham Corporation v Charles Berabo & Sons* [1934], it was possible to deduce the proposition that where a notice or demand was required by Parliament to contain a particular content and it was plain that only a notice with that content was declared to have the effect of a statutory notice, a document which failed to include that content or made some significant omission from the content required did not qualify as a notice under the statute. It was not a defective or invalid notice which could be cured by amendment or otherwise; it simply never began to have any statutory force or effect. The court found that the document in this case was not therefore a notice under the Act.

Not only must a notice, when required, be specific it must also indicate clearly the point at which the recipient can be sure that he or she has complied with all its requirements. For example in *Metallic Protectives Limited v Secretary of State for the Environment* [1975] the Divisional Court found an enforcement notice, alleging a breach of a planning condition, to be imprecise. The notice required that no nuisance should be caused to the residential properties in the area by reason of the emission of noise, vibration, smoke, smell, fumes, soot, ash, dust or grit. It went on to require the occupier in question to install satisfactory sound proofing of a compressor and to take all possible action to minimise the effects created by the use of acrylic paint. The notice was held to be imprecise as it did not indicate a point at which the occupiers could be certain that they had satisfied its requirements and so be sure that they were free of the notice. It was a kind of continuous or open-ended obligation and so it did not specify the steps required to be taken in order to remedy the breach. It was a nullity and incapable of amendment. To use the words of Upjohn LJ in *Miller-Mead v Minister of Housing and Local Government* [1963] (at p.232) the notices should be set out with sufficient particularity to meet the requirements of fairness and to enable the recipient to know what they had done wrong and what they must do to remedy it.

In the case of *Strathclyde Regional Council v Tudhope* (1983), the City of Glasgow District Council had served a notice under the provisions of section 58 of the Control of Pollution Act 1974 on Strathclyde Regional Council. After describing the relevant area, the notice stated:

> The City of Glasgow District Council hereby give you notice that … there exists, or is likely to occur or recur, noise amounting to a nuisance within the meaning of the Control of Pollution Act 1974 viz. Noise from road breaking operations is excessive in that the best practicable means are not being taken to minimise noise and you as the persons responsible are required to:
>
> 1. Fit all pneumatic breakers in use with effective exhaust silencers
> 2. Fit all pneumatic breakers in use with dampened tool bits
> 3. Use only suitably silenced compressors
> 4. Keep all side panels on air compressors closed whilst the said compressors are in operation

> 5. Ensure that there are no leaks in air lines used in conjunction with the said compressors or breakers,

with immediate effect from this date.

Strathclyde Regional Council were subsequently charged at the instance of the Procurator Fiscal, James Tudhope. The appellants argued, amongst other matters, that the notice was void for uncertainty, the use of words such as 'suitable' and 'effective' being the basis for this challenge. In the High Court of Justiciary Lord Wheatley found that there was no uncertainty, what was being required was clearly specified in the notice. He found that the first three requirements were in effect saying that to secure the desired noise abatement all pneumatic breakers in use had to be fitted with effective exhaust silencers and dampened tool bits and all compressors in use had to be suitably silenced. If they were not so equipped they were not to be used until they were. He saw nothing vague or uncertain about this. The fourth and fifth requirements also appeared to be perfectly straightforward and did not disclose an impracticability in the issuing of instructions for their being carried out.

In the case of *Network Housing Association v Westminster City Council* [1995], the council served a notice pursuant to section 79(1)(g) of the Environmental Protection Act 1990. The notice required the appellants to:

> provide suitable and effective sound insulation in the void between flats D and C, so as to provide a level of airborne sound insulation (measured as DnTw in accordance with BS2750 part 4 (1980) and BS5821 part 1 (1984)) of not less than 42 dB or carry out such other works as will achieve the above required degree of airborne sound insulation between flats D and C. In improving the airborne sound insulation take all reasonable steps to ensure that no degradation occurs in the existing level of structure-borne sound insulation.

The appellant challenged the notice in the magistrates court, asking whether the notice adequately specified the works required. Buckley J found that, bearing in mind the risk of exposure to penal sanctions for non-compliance, it was essential that the recipient of such a notice should be told clearly what works were required to be carried out. He accepted that in some obvious cases a notice requiring little more than that a certain result be achieved might suffice. But he went on to say that in the circumstances of this case, which involved a notoriously difficult question of sound levels and nuisance, he considered that the respondents should have made up their own minds as to the work required and stated it in the notice and for that reason this notice fell short of the minimum legal requirement to convey to its recipient clearly what it had to do.

Simon-Brown LJ found that the notice left it to the appellants to decide what was an appropriate solution to the problem. He found such an option to be a poisoned chalice because if the recipients' choice of work failed to produce the required result (the stipulated degree of sound reduction) they would then be bound to attempt further works. The notice, in short, instead of stating the specific works to be done or action to be taken, contemplated instead the possible need to take a succession of works. That, in his words, would not do.

There is therefore a requirement on any Environmental Health Officer proposing to serve a notice to ensure that if the statutory provision requires works to be specified they must be so specified. This obligation can cause a number of practical problems and begs the question as to what the limits of such a requirement are. It may for example be practically impossible to specify fully all works which may be required. In *Church of Our Lady of Hal v London Borough of Camden* (1981) the local authority served a repairs notice under s.9(1A) of the Housing Act 1957 on a property which required substantial repairs to be carried out. The works were contained in a lengthy schedule attached to the notice. This contained a number of paragraphs relating to the work to be carried out and would be of a form very familiar to Environmental Health Officers. With regard to works to the roof the schedule read:

> Thoroughly overhaul the main roof. Clean out and examine the parapet gutters and replace or renew as necessary. Take off and renew all defective, slipped and missing slates. Hack out all cracked and missing fillets abutting the roof and properly renew. Properly make good the roof timbers as necessary. Renew any defective soakers or flashings.

The court felt that the bulk of that paragraph was simply further and better particulars of the principal expression 'thoroughly overhaul the main roof'.

On matters relating to pipework the schedule stated: 'Properly examine waste pipe, soil vent pipe and rainwater pipe. Renew where necessary and make good to all joints, using appropriate jointing materials.' Again Oliver LJ felt that this wording indicated with sufficient precision the defects that had to be looked for and rectified. With regard to windows in the property the notice required the recipient to:

> overhaul the window woodwork and sashes, renew any defective and rotted timber to the sashes, frames and sills. Renew any badly worn and defective sashes, broken sash cords, catches, glass and beading. Ease and adjust the sashes and weights as necessary and leave in sound working order on completion.

Oliver LJ again felt such wording made it perfectly clear what the owner had to do. He had to inspect the window, make sure it was in working order and replace any defective or rotted timber which was found on his inspection. In examining this case the court accepted a number of propositions with regard to the wording of such notices which are of general applicability. Firstly such a notice should contain sufficient information in its schedule to enable the building owner to have the work costed out by a reasonably competent builder. Secondly there will generally be no objection to the use of the words 'as necessary or where necessary' in describing works, where the work described is either: (a) ancillary to or consequential on other work; or (b) it is a matter of judgement for the builder to decide what work will be necessary to be carried out in relation to defects which have been pointed out.

Only in very rare cases where the content of the notice is so vague that the owner cannot know what the cost of repairs would be with regard to the major requirements of the notice, should a court exercise its power to quash a notice without evidence. The court in this case found that a notice does not have to set out the exact location, the exact area, the exact nature of every piece of work that had to be done. The local authority was obliged to show the owner with reasonable precision what he has to do, but not to dot every 'i' and cross every 't'.

Such an approach is in line with the earlier decision in *Cohen v West Ham* [1933]. Here a notice under s.17 of the Housing Act 1930 contained a requirement to: 'take down and rebuild the front, back main and back addition walls where necessary'. Maugham J found that when dealing with work of this nature a notice should specify with reasonable clearness what portion of the wall it is necessary to begin to deal with for the purpose of putting right the defect. In such cases it would be possible to specify that it was necessary to rebuild the wall at a particular angle of the building, or the wall above the string course, or the wall from the ground level, or the eastern portion of the building, or the part of the building where there is a bulge in the wall, or that part of the building which is out of the true to a particular extent. All those things could be specified. A notice to rebuild such part of a wall as may be necessary or to rebuild a wall where necessary leaves it open to a recipient to say, on examination with the assistance of a builder or an architect, that it is not necessary to take down and rebuild any part of the wall and that another step will be sufficient. Such a notice would be wholly bad for want of precision if it were to be construed as a notice to take down a part of the wall without specifying what part.

In *R v Fenny Stratford Justices, ex p. Watney Mann (Midlands) Ltd* [1976], Watney Mann owned a pub situated in a building which also contained residential flats. Residents of the flats complained of noise made by a juke box in the public house and applied under s.99 of the Public Health Act 1936 (see now s.82 Environmental Protection Act 1990) to the justices to make a nuisance order requiring Watneys to abate the nuisance. The justices made the order, and in it specified that the level of noise should not exceed 70 decibels. Watney sought to quash that part of the order specifying the noise level on the grounds that the justices had no power to make such a requirement and that it was void for uncertainty. It was held that a court order specifying that a level of noise should not exceed 70 decibels was void by reason of its failure to specify the exact point at which the noise was measured. The order was so imprecise as to be void for uncertainty.

It would be wrong to interpret these findings as the courts exercising an overly pedantic degree of supervision. Certainly the courts are aware of the need to ensure that notices retain their practical use and are not overly inclined to be misled by artificial attempts to attack them. In *Stevenage BC v Wilson* (1991) a notice was served under s.58 of the Control of Pollution Act 1974 (see now s.80 Environmental Protection Act 1990) prohibiting the recurrence of a nuisance consisting of the playing of recorded music 'at an excessive high volume at 13 Tintern Close'. The notice further required the recipients to: 'take all steps necessary to prevent the disturbance by noise of the inhabitants of the neighbourhood and in particular to ensure that the amplified music emanating from the dwelling is not clearly audible outside the house or in the adjoining dwellings'.

Subsequently the authority laid an information alleging breach of the requirements of the notice by the playing of loud music in the garden. On hearing the case magistrates dismissed the information, finding that the notice was not clear on the point that it was intended to cover all land within the boundaries of the property as well as the dwelling itself. They appeared unconvinced that dwelling included the garden. The prosecution appealed by way of case stated. On appeal the court found that the reference to '13 Tintern Close' necessarily included the garden as well as the house and that the word 'Dwelling', when viewed in context, was also apt to include the garden. This is a reassuringly sound finding. It would be nonsensical to take a narrow and restrictive view of the meaning of the words used when having regard to the purposes of the statutory provision.

## 8.11 THE APPROPRIATE USE OF ALL INVESTIGATIVE OPTIONS

It is clear that to be able to draft a notice officers must be prepared to use all investigative options to inform firstly their decision to serve the notice and secondly what to include in the notice specification. To do otherwise is to run a number of risks: the risk of serving a notice improperly, or the risks of failing to specify with sufficient particularity or of imposing conditions which are more onerous than truly required required. Environmental Health Officers are given extensive investigative powers under a wide range of statutory provisions. Officers should not view these as optional. The question ought not to be 'which one of these shall I use' but 'which of these is it not appropriate to use'. By considering all investigative options, and then positively rejecting those which are inappropriate, an officer can be sure that he will not overlook invaluable information. This information may be the key to the correct drafting of the notice, as it may identify what the central issue in the problem is and thereby help the officer to decide on the appropriate specific remedial works to include in the notice, rather than some vague or blanket requirement. It may also reveal some defence that makes the service of the notice of questionable validity, for example a best practicable means defence to a notice under s.80 of the Environmental Protection Act 1990.

It is certainly arguable that an officer who fails to exercise all investigative options cannot carry out a proper risk assessment which may implicitly or explicitly be a condition precedent for the service of a notice. This has been considered by an Industrial Tribunal with regard to the service of notices under the Health and Safety at Work etc. Act 1974.[7] In this case Kwik-Save appealed against an improvement notice issued by an EHO requiring the replacement of stock knives with safety knives. Kwik-Save contended that the EHO could not have competently identified the appropriate safety improvements necessary because she had not taken as much care over risk assessment as the company. The tribunal recorded that the officer had made insufficient use of her investigative powers under the Health and Safety at Work etc. Act 1974. It found that the officer should have carried out a proper risk assessment in order to form an opinion that the company was in breach of its duty. The only way to form such an opinion, which can then be defended, is to use all appropriate investigative powers to the extent necessary to act in an informed manner. It is only in this way that the decision to serve this and other similar types of notice can hope to withstand such a challenge.

### 8.12 CONFLICTS ON THE FACE OF THE NOTICE

Conflicts on the face of the notice may easily occur due to basic errors when composing the notice. Many pre-prepared notices contain a number of elements which, according to circumstances, require to be deleted before the notice is served. It is all to easy under pressure of time to retain or omit an element which will subsequently result in a conflict on the face of the notice. For example it is common to use the prescribed forms for repairs notices under s.189 of the Housing Act 1985 in their most basic form, i.e. reproduced directly from the Housing (Prescribed Forms) (No 2) Regulations 1990. In this form a notice will require a number of elements to be deleted before the general form becomes specific to the case in hand. Carelessness at this stage may for example result in the notice referring to group repair works where none are in fact underway, or it may refer to the person managing when the recipient should be the person having control. Such errors are avoidable but if permitted to occur will almost certainly render the notice voidable if not void.

It may be that in drafting a notice officers fail to recognise that the notice reveals a potential conflict. Where an abatement notice under the Environmental Protection Act 1990 requires works to be carried out, these must as already discussed be specified. When all that is required is to stop a thing being done, then, following *McGillivery v Stephenson* [1950], the mere prohibition of that thing is sufficient. This case involved a notice which required the cessation of pig keeping without giving an alternative method of abating the nuisance. So when serving a notice under s.80 of the Environmental Protection Act 1990 which requires a person simply to stop an activity this may be satisfactory. However a notice under these provisions is regulated by the Statutory Nuisance (Appeals) Regulations 1995 which deal with appeals under section 80(3) of the EPA 1990. Regulation 2(2) details the grounds on which a person served with an abatement notice may appeal. The effect of entering an appeal may, subject to circumstances, be to suspend the operation of the notice. A notice will for example be suspended if compliance with the notice involves any person in expenditure on the carrying out of works before the hearing of the appeal. It will not be suspended if, inter alia, the local authority considers that the expenditure incurred in carrying out works in compliance would not be disproportionate to the public benefit resulting from compliance prior to the hearing of the appeal. Were the notice to contain a mere prohibition or specify an outcome, but not the works required to achieve that outcome, then, if that notice was not suspended pending an appeal, it may reveal a conflict on its face. Essentially a local authority might find it difficult to claim that costs of work are not disproportionate to the benefit where they fail to specify those works. Non-specific works cannot be quantified or costed and therefore their benefit cannot be weighed against cost.

On the point of a notice merely specifying an outcome is *Myatt v Teignbridge DC* [1995]. Here the abatement notice in question specified that the recipient should 'cease the keeping of dogs'. Butler-Sloss LJ felt that such wording could in many cases be unsatisfactory. The recipient of such a notice should know what is wrong but the wording used here did not disclose this. The judge pointed out that the simple keeping of a dog or dogs would not amount to a nuisance. Thus the notice was clear about what had to be done but was unclear about what was wrong. Words such as a nuisance 'arising from the keeping of dogs which are barking loudly at night' would have been clearer as to what the remedy specified was seeking to deal with. Despite these observations the appellant here lost her case as the court felt that any person keeping 17 dogs in the circumstances disclosed must have known what was objectionable. On the question of this type of notice s.80(1)(a) of the Environmental Protection Act 1990 provides that an abatement notice may be served requiring the abatement of the nuisance or prohibiting or restricting its occurrence or recurrence. On the face of it it is therefore arguable that a single notice which purports to both abate a nuisance and restrict its recurrence may go beyond the provisions of the Act. If this is accepted it may be more appropriate to serve two separate notices, one to abate the nuisance, the other to prevent recurrence.

The problem is not unique to Environmental Protection Act notices. A similar problem can arise under housing legislation. A number of authorities, based on guidance from the Department of the Environment, serve notices under ss.189 and 190 of the Housing Act 1985 at the same time. This is despite the fact that one says that the house is unfit and the other says it is fit. There exist some complex and not entirely persuasive arguments in favour of this course of action, but nevertheless on the face of it the notices are in conflict and this should be positively acknowledged by the officer serving the notice. It is only by thoroughly examining notices for such conflicts that an Environmental Health Officer can expect to be able to defend his actions should his notices be tested before the courts.

### 8.13 USE OF SCHEDULES

It is common practice for notices to contain schedules for the work they require. These will take the form of a list of works which the notice requires to be done. In practice there can be little objection to this approach but it does have certain consequences. The first point to bear in mind when using such schedules is that if there is an error or defect in the notice this may have the effect of rendering the entire notice, including all items in the schedule, void. Therefore all items on the schedule will be lost. This may result in quite substantial amounts of work being rendered useless by a relatively minor error. Though legally it is in some cases more advisable, it is hardly practical to serve a separate notice for each contravention however minor. Thus if schedules are to be used certain disciplines should be observed which ought to help to maximise the benefit accruing from the use of those schedules whilst minimising their potential problems.

(a) Each clause should be numbered; this makes for easier identification of relevant matters in any subsequent examination of the notice.

(b) Ensure the recipient is clear on the number of pages making up the schedule that ought to be attached to the notice and that each page of the schedule carries the address of the property concerned.

(c) Where there are specific requirements regarding the form and content of a notice these must be carried over into the use of schedules. For example notices served under s.10 of the Food Safety Act 1990 require that the wording of the notice should be clear and easily understood. Any improvement notice must include details of the regulation contravened and the reason for the opinion of the authorised officer that there has been a contravention. Thus in its guidance on the drafting of improvement notices LACOTS (Food Safety Act 1990 – Guidance on the Drafting of Improvement Notices, LACOTS FG 4 952 1 22 November 1995) correctly points out that when using schedules it is important to take care to ensure that for each of the contraventions listed the relevant regulation, schedule, Chapter, requirement, opinion of the inspector and remedial measures are easily identifiable.

(d) If notes are required to be sent with a notice ensure that they are printed on the back of the notice or that there is some positive means of recording that they and all pages of a schedule have been sent with the notice.

(e) Where a number of rooms are included in a specification of works the use of sub-headings to identify the room is appropriate. Such headings should avoid referring to the room by its use, e.g. kitchen, but should instead refer to it by location, e.g. ground floor rear room. Where such is to be used a clear statement regarding method of orientation should be included on the notice.

(f) Avoid clustering works of the same type together, e.g. window repairs or renewal. Each window repair should have a separate clause relating to the works which are required. This will help to provide as clear a statement as possible of the works required and make it less likely that any element will be overlooked either by the recipient or by the officer on reinspection.

## 8.14 TIME PERIOD

Where a statute provides that an authority may serve a notice requiring work to be done it will often, but not inevitably, specify the minimum period such a notice may carry; otherwise it will often offer no guidance on the appropriate time to be stipulated within the notice. In all cases where no statutory time period is stipulated the basic rule is that the

Ref. No..............

**Page 1 of ....**

**SCHEDULE**
Address of Premises............................................................................

Statutory Provision..........................................................................

Notice sent to..................................................................................
......................................................................................................

Date of Notice.................................................................................


This schedule consists of ........ items.

Where relevant the method of orientation used for items is (from the outside of the property facing the front of the building (name of street)/from inside of the building facing the front (name of street))

**Item**

Form 8.2 Schedule

minimum period, beyond any statutory minimum, that may be specified in any notice must be a period of time within which the work can reasonably be completed, having regard to the nature of the work and all the circumstances (*Thomas v Nokes* (1894)). Therefore the least time that can be fairly and reasonably be given for the completion of any work is the shortest period that may be inserted in the notice. An officer will of course have the discretion to allow a further time for the completion of the work, but not to fix a shorter time. The rationale behind this is quite clear. Statutory notices will not permit the insertion of a time period which will inevitably result in the recipient committing an offence of non-compliance. The courts will, even in the absence of a specific statutory provision allowing for an appeal relating to an unreasonably short period of time, retain a supervisory jurisdiction to examine notices where such is the case. The problem for anyone proposing to serve a notice is deciding what is a reasonable period of time. Food Safety Act 1990 Codes of Practice offer relevant guidance on what considerations are material in deciding on an appropriate time to specify in a notice and this, with appropriate amendment, is applicable to all notices, not simply only those under the Food Safety Act. The guidance suggests that the period given for completion of the work should be a realistic one. With this in mind the following factors should be taken into consideration before a time limit is set:

(a)  the nature of the problem;
(b)  the risk to health;
(c)  the availability of solutions.

In addition to these an officer's previous experience of similar matters will help to inform the decision as to what is an appropriate period of time. In all instances the decision to insert a period of time on a notice should be the result of positive thought by the officer concerned rather than simply putting in the 'standard' time. This should ensure that if subsequently challenged the officer will be able to easily account for, and thus defend, the time period inserted on any notice he or she may serve. The decision regarding the time period is of course that of the officer and does not require the recipient's agreement. There is however no reason why the period set for compliance cannot have been agreed with the recipient before the notice has been served. It would be inadvisable for officers to get into haggling over the period of time, but a time which is both reasonable and accepted by the recipient of a notice will, should there be the need for subsequent enforcement action, be viewed as much more acceptable by a court and be less likely to be treated unsympathetically.

Where schedules are attached to a notice there should not be inserted differing compliance dates for different elements on the schedule. This approach would act so as to make the notice invalid. Either a single date is to be inserted or separate notices should be served.

Though notices served under s.80(1) of the Environmental Protection Act 1990 will often require works to be done, sometimes a simple prohibition will suffice which would allow the recipient to act on the notice immediately. In such instances it may not be necessary to specify a time limit for compliance, and the notice will take immediate effect; see *Strathclyde Regional Council v Tudhope* (1983).

Notices served under these provisions can impose a prohibition on matters recurring (s.80(1)(a)). Here there arises the question as to whether there is an implicit time limit beyond which notices containing such a prohibition may be considered expired. The case of *R v Birmingham JJ ex parte Guppy* (1988) examined this point with regard to similar provisions found in s.58 of the Control of Pollution Act 1974. The substance of the case involved an application for Judicial Review to quash the decision of the magistrates convicting Guppy of three offences under section 58. The Court of Appeal held that a notice under section 58, which included a requirement prohibiting the recurrence of a nuisance, was not subject to a time limit for compliance. The prohibition took effect as soon as the requirement to cease causing the nuisance was complied with and the prohibition remained in force indefinitely. This indefinite prohibition remains even though s.58 has now been repealed. The House of Lords in *Aitken v South Hams DC* (1994) considered the status of such notices following the passing of the Environmental Protection Act 1990. The offence under section 58(4), of which Mrs Aitken had been found guilty, had been in respect of a notice served on 25 November 1983 alleging that she had allowed noise nuisance from barking dogs to recur between 30 August and 15 October 1991, after the Environmental Protection Act 1990 had come into force. The issue before the court was whether a person who contravened without reasonable excuse a notice served prior to 1 January 1991 was guilty of an offence. The court held that notices dealing with noise nuisance under section 58 of the Control of Pollution Act 1974 have to be complied with even though that section has now been repealed.

### 8.15 ADEQUATE AUTHORITY

Section 101 of the Local Government Act 1972 Act gave local authorities the general power to discharge any of their functions through their officers and it is now common for such officers to have specific areas of responsibility, supervised generally via the local authority committee system. One such common area of delegated responsibility is the signing of notices. Section 234 of the Local Government Act 1972, which deals with the authentication of documents, requires that any notice, order or other document which a local authority are authorised or required by or under any enactment to give, make or issue may be signed on behalf of the authority by the proper officer of the authority. A proper officer would be an officer who possesses a power to sign, delegated to him under s.101. All notices must therefore be signed by an authorised officer. This signature need not however be the actual signature applied by the officer concerned. Section 234(2) of the Local Government Act 1972 stipulates that any document purporting to bear the signature of the proper officer of the authority shall be deemed, until the contrary is proved, to have been duly given, made or issued by the authority of the local authority. In this subsection the word 'signature' includes a facsimile of a signature by whatever process it is reproduced. Thus it is that the common practice of the issuing of a notice by a subordinate using a facsimile imprint of an authorised officer's signature is valid and a proper procedure (*Fitzpatrick v Sec. of State for the Environment* (1990), *Plymouth Corpn. v Hurrel* [1967]). It would appear that this may extend to the computer reproduction of a signature which is now possible using easily available standard technology. The signing of notices should not be undertaken lightly nor should the power to sign notices be too widely delegated. Guidance in the Codes of Practice under the Food Safety Act 1990, though only relating to food safety enforcement, again offers advice which is equally applicable to all notices, not simply those under the Food Act. The advice suggests that notices should be signed only by officers authorised to do so by an enforcement authority and that such persons should be qualified officers with experience of enforcement. It recommends that an authority, before authorising an officer to issue notices, should be satisfied that the officer is competent to do so and possesses experience of enforcement. There can be little objection to this being the approach to be adopted for all enforcement notices.

It is clearly correct that only those authorised to sign notices should do so and though it has been held that the subsequent ratification by the authority, if given before legal proceedings are commenced, may be sufficient to give a notice legal effect (*Warwick Rural District Council v Miller-Mead* [1962]) this is a dangerous approach to adopt. In *Webb v Ipswich BC* [1989] the limits of such retrospective validation were made clear. A purported control order under s.379(1) of the Housing Act 1985, which was made by an officer of the council who did not have the appropriate authority, could not be retrospectively validated by the council sub-committee even though it did have the authority to make such an order. Lord Justice Croom-Johnson distinguished this matter from *Miller-Mead* by saying that ratification could not be made if an individual's legal rights had been affected by the valid act. The case was not one of retrospective correction of a formality which had not been observed, nor was it the straightforward ratification of the act of an authorised agent. It was an attempt to put right something which had been carried out by an unauthorised agent in the first place. This is very much in line with the approach which may be traced at least as far back as *St Leonard's Vestry v Holmes* (1885). In this case a notice was served by a sanitary inspector without authority. Later a second notice was served by the same person but after consultation with some members of a sub-committee. No consideration was given by the Vestry before the work was done. After completion of the work the actions of the inspector and the sub-committee were approved by resolution of the Vestry. The Vestry then sought to recover the costs of having carried out the works in default of the owner of the premises. It failed. In the words of Day J: 'it is important that the vestry should exercise a discretion in each case, and it is not enough that the inspector does what he pleases, and then relies on his acts being afterwards approved by the vestry.'

### 8.16 SERVICE OF NOTICES

For all of the statutes which allow for the service of statutory notices some method of service is specified (Public Health Act 1936, s.285; Building Act 1984, s.94; Housing Act 1985, s.160(2), s.571, s.617; Local Government Act 1972, s.233; Environmental Protection Act 1990, s.160).

As an example s. 160 of the Environmental Protection Act 1990 at subsection (4) states that:

> For the purposes of this section and of section 7 of the Interpretation Act 1978 (service of documents by post) in its application to this section, the proper address of any person on or to whom any such notice is to be served or given shall be his last known address, except that
>
> (a) in the case of a body corporate or their secretary or clerk, it shall be the address of the registered or principal office of that body;
> (b) in the case of a partnership or person having control or the management of the partnership business, it shall be the principal office of the partnership;
>
> and for the purposes of this subsection the principal office of a company registered outside the United Kingdom or of a partner-

ship carrying on business outside the United Kingdom shall be
their principal office within the United Kingdom.

It is clear from this what the proper address for the purposes of s.7 of the
1978 Act is. Section 7 of the Interpretation Act 1978 provides that 'where
an Act authorises or requires any document to be served by post then,
unless the contrary intention appears, the service is deemed to be
effected by properly addressing, pre-paying and posting a letter containing the document and, unless the contrary is proved, to have been
effected at the time at which the letter would be delivered in the ordinary course of post.'

For most of the statutes provision is made for service by more than
one means. Ordinary post is the most obvious means of service and can
constitute good service. However where legal proceedings are anticipated, and this must be in all instances when a notice is served, ordinary
post, because of the evidential difficulties of proving service, can be
unsatisfactory.

An alternative method is to serve by registered post or more commonly recorded delivery. These methods when coupled with an Advice
of Delivery card can be used to establish good service. Such a card contains details of the delivery of the letters. It contains the number of the
item, the date of posting, the name of the person to whom it was delivered or who called for it, the date of that delivery and the signature of
the recipient and the signature of the post office employee. In *Chiswell v
Griffen Land and Estate* [1975], Lord Justice Megaw said that if it is proved,
in the event of a dispute, that a notice was sent by recorded delivery it
does not matter that the recorded delivery letter may not have been
received by the recipient. In *Lex Service plc v Johns* (1989) the court found
that where a notice, here under the Landlord and Tenant Act 1954, was
sent by recorded delivery and was recorded as having been received at
the address, a simple denial by the proposed recipient of the notice did
not have the effect, for the purpose of s.7 of the Interpretation Act 1978,
that the notice had not been served within the provisions of that section.
Lord Justice Glidewell said that what was needed to prove the contrary,
in s.7, was positive evidence that the document had been returned to
sender, or in the case of a registered or recorded delivery letter that there
was no acknowledgement of its having been received or if there was evidence of receipt by someone, that the receiver was not the person on
whom the document was served and the receiver had not drawn it to
the attention of the other. A mere assertion by the intended recipient
that the document had not been received by him did not suffice.

When any criminal proceedings are anticipated then there is no doubt
that personal service is the best method, though it has the disadvantage
of being expensive, time consuming and, because the officer may be

known to the recipient, it may put the officer at some risk where feelings are running high. In the case of *Rushmoor BC v Reynolds* (1990) a Technical Officer of the authority served a notice under s.16 of the Local Government (Miscellaneous Provisions) Act 1976 requesting details regarding a property at 9 The Warron, Aldershot. The notice was served by posting it through the letter box of 22 Anglesey Road, Aldershot. No 22 was a house in multiple occupation and the respondent had previously had difficulty in mail reaching him. The notice did not reach Reynolds and he was subsequently charged under s.16 with failing to respond to the notice. On hearing the case magistrates acquitted him of the charge. Though the notice had been served in accordance with s.233 of the Local Government Act 1972, as it had not been handed to the respondent personally he could rebut the evidence of service and had done so on the balance of probabilities. Rushmoor Borough Council appealed by way of case stated. The Queens Bench Division allowed the appeal and remitted the case back with a direction to convict. It was found that a notice served by hand or by post under s.16 of the 1976 Act or by any of the means specified in s.233 of the 1972 Act gave rise to an irrebuttable presumption that it had reached the person to whom it was directed, therefore no defence was available. On the same point, in the case of *London Borough of Lambeth v Mullings* (1990), it was held that the service of an abatement notice under s.58 of the Control of Pollution Act 1974 (see now s.80 Environmental Protection Act 1990) could be through the letter box of the subject premises.

Because it is to be anticipated that notices may result in a case before a magistrates court then service in accordance with the Magistrates' Courts Rules 1981 would usually be taken to be an adequate method. Rule 99 states that service of a summons is effected by delivering it to the person to whom it is directed or by leaving it for him with some person at his last known or usual place of abode or by sending it to him by letter addressed to him at his last known or usual place of abode. Rule 67 states that for the service of any process or other document required or authorised to be served, the proper addressing, pre-paying and posting or registration for the purpose of service of a letter containing such a document, and the place, date, time of posting or registration of any such letter, may be proved in any proceedings before a magistrates court by a document purporting to be a certificate signed by the person by whom the service was effected or the letter posted or registered. Whenever any notice is served a certificate of service should be included with or attached to the notice as a record of the means by which service was effected. An example of such a certificate is reproduced.

## CERTIFICATE OF SERVICE BY POST (MC RULES 1981, rr 67, 99(6))

I ..............................of Anytown City Council, hereby certify that I served
.........................................................................................................................
with the Notice of which this is a true copy, by sending the said Notice by
the recorded delivery service to them in a prepaid letter posted at the
POST OFFICE situate in .........................................................................
at ........... in the morning/afternoon on the ........... day of ..............and
addressed to.....................................................................................................
being their Registered Office/home address

Signed
.........................................................................................................................

Dated the .............. day of ..........19........

**Form 8.3** Certificate of service by post

## CERTIFICATE OF SERVICE BY HAND

I .............................. of Anytown City Council, hereby certify that I served
.........................................................................................................................
with the Notice of which this is a true copy, by delivering it by hand to
.........................................................................................................................
at ........... in the morning/afternoon on the ........... day of .................. and
addressed to ...................................................................................................

Signed
.........................................................................................................................

Dated the .............. day of ..........19........

**Form 8.4** Certificate of service by hand

## 8.17 KEY POINTS IN THE DRAFTING OF NOTICES

What points can be made to minimise likely problems with the drafting of notices.

1. A notice must correctly identify the recipient as a person who to whom such a notice may be sent.
2. That person's full name must be on the notice. Avoid using only initials. If the notice is to a company the full name of the company must be used.
3. The full postal address must be used. For a company that is their registered office.
4. The notice must be in writing.
5. The notice must use an adequate standard of English.
6. A notice must comply with the relevant statutory requirement.
7. Any prescribed format must be adhered to.
8. If a prescribed format is followed ensure all irrelevant elements are deleted.
9. If schedules are used ensure that they adhere to the prescribed format
10. Number all paragraphs in any schedule to ensure clarity and ease of identification.
11. Number all pages of a schedule and indicate the total number of pages attached to the notice.
12. Ensure all pages in a schedule carry the address of the property in question.
13. Ensure all pages in a schedule identify the statutory provision under which the notice is served.
14. The notice must be signed by an authorised officer or by some other means have an authorised signature affixed.
15. The notices should be set out with sufficient particularity to meet the requirements of reasonableness and fairness.
16. The notice must be phrased so as to enable the recipient to know what they had done wrong and what they must do to remedy it. It must be precise, technically competent and not onerous.
17. A notice should contain sufficient information to enable the recipient, where appropriate, to have the work costed out by a reasonably competent builder.
18. There is no objection to the use of the words 'as necessary' in describing works, where the work described is either ancillary to or consequential on other work, or it is a matter of judgement for a builder to decide what work will be necessary to be carried out in relation to defects which have been pointed out.

19. Only in very rare cases, where the content of the notice is so vague that the owner cannot know what the cost of repairs would be with regard to the major requirements of the notice, should a court exercise its power to quash a notice without evidence.
20. A local authority must show the owner with reasonable precision what he has to do, but to dot every 'i' and cross every 't' is recognised as impracticable.
21. What is required is a necessary and reasonable degree of particularity.
22. It is not vital that a notice should have to set out the exact location, the exact area, the exact nature of every piece of work that has to be done.
23. A notice to rebuild such part of a wall as may be necessary, or to rebuild a wall where necessary, leaves it open to the recipient to say, on examination with the assistance of a builder or an architect, that it is not necessary to take down and rebuild any part of the wall and that another step will be sufficient.
24. A notice should specify with reasonable clearness what portion of the property it is necessary to begin to deal with for the purpose of putting right the defect, e.g. above the string course, or the wall from the ground level, or the eastern portion of the building, or the part of the building where there is a bulge in the wall. A notice may be bad for want of precision if it could be construed as a notice to, for example, take down a part of the wall in any event without specifying what part.
25. The least time that can be fairly and reasonably allotted for the completion of the work is the least time to be inserted in the notice; there is a discretion to allow a further time for the completion of the work, but not to fix a shorter time.
26. The notice must not reveal any conflicts on its face.
27. Proof of service must be ensured.

## NOTES

1  Hutter, BM (1988) The Reasonable Arm of the Law: The Law Enforcement Procedures of Environmental Health Officers, *Oxford Socio-Legal Studies*, Clarendon Press.
2  Hawkins, K, 1984, *Environment and Enforcement: Regulation and the social definition of pollution*, Clarendon Press, p. 179.
3  Robens Committee (Cmnd. 5034 (1972)).
4  Health and Safety at Work etc. Act 1974. Ref. No.: HSC(G) 3, 24 January 1996.
5  Housing Act 1996 and Housing Grants Construction and Regeneration Act 1996.
6  Review of new health and safety enforcement procedures, HSE Books.
7  Health and Safety Information Bulletin 213, September 1993. Industrial Tribunal Nottingham, 11 February 1993.

# Pathway of a Case to Court

# 9

## 9.1 FILE PREPARATION

Before a decision to prosecute is taken there will have been an investigation of the offence. This will initially have been by means of obtaining statements from any complainant and witnesses and by inspection of premises and articles or substances. Later there may have been interviews undertaken having regard to the requirements of the Police and Criminal Evidence Act 1984 and the relevant codes of practice. These interviews may have been taped or a written record may have been made. Where appropriate exhibits will have been obtained and some form of prosecution file assembled. Currently there exists for Environmental Health professionals no national standards as to what such a prosecution file should contain. Each authority therefore operates according to its own rules and standards. There does exist however some guidance as to what such a file ought to contain. In 1990 the Joint Agency Working Group on Pre-trial Issues made a number of recommendations in relation to police prosecution files which, with appropriate amendments, could be suitable for the work of the Environmental Health Officer. The group comprised representatives from the Lord Chancellor's Department, the Crown Prosecution Service, the police, the Justices Clerks Society and the Home Office. As a result common national standards for prosecution files now exist and police forces now use common national forms on all files. These forms include:

- File front sheets
- File content check lists
- Witness lists
- Defendant detail forms
- Witness non-availability forms
- Confidential information forms

- Criminal Justice Act statements
- Exhibit list
- Record of Taped Interview forms.

Recommendations regarding the preparation and submission of police files to the Crown Prosecution Service have been incorporated into a Manual of Guidance which is issued to all forces. Liaison with their local police force on this is likely to prove useful to Environmental Health departments. Since November 1995, following a recommendation of the Pre-Trial Issues Steering Group, it has been seen to be appropriate to provide shorter case files for cases likely to result in a guilty plea. Such files will only contain:

- Key witness statements, establishing each element of the offence
- A guide to the prosecution case. This would indicate where the evidence to prove the elements of the offence are to be found.

Additionally short descriptive notes will replace records of taped interviews. Such notes will contain brief summaries of any admissions made by an accused, any aggravating or mitigating features and any other relevant material. It may be that Environmental Health Officers should consider this as an additional useful model from which to learn. Such a file would then be reviewed by senior officers and lawyers before a decision to institute proceedings is taken.

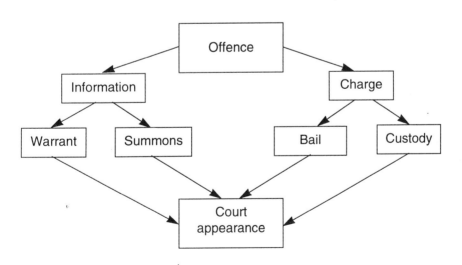

**Figure 9.1** Pathway of a case to court

## 9.2 INFORMATION AND SUMMONS

### 9.2.1 THE PROSECUTION OPTION

There may be a view that for an Environmental Health Officer to find himself in court is to accept that he has somehow failed; that his role as educator and persuader has proved ineffectual. This is wrong. It is Thomas Fuller who is credited with saying that law cannot persuade where it cannot punish and this surely holds no more true than in the work of the Environmental Health Officer.

An appearance in a court of law is an inevitable and necessary part of the work. The public, who invest their confidence in the Environmental Health Officer, do so in the anticipation that, when it is appropriate, the officer will not shrink from instituting proceedings. Earlier there has been an exploration of the circumstances when prosecution may be 'appropriate'. If this is accepted, then a prosecution can be seen merely as another facet of the work of the Environmental Health Officer. It has its role and its purpose, both for the matter in hand and for the more general protection of health and the environment.

The pathway of a case to a court is relatively straightforward. An individual or company will appear before a court either because he has been charged with an offence or because a summons has been issued. The charging of a person will follow an arrest; this is a power not available to the Environmental Health Officer and therefore he is most likely to deal with appearances attributable to the latter process. It is this process which will be examined in more detail. The means by which a person is called before the magistrates court to answer criminal matters is by a summons and this is obtained by the laying of an information before the court.

### 9.2.2 LAYING INFORMATION

An information is essentially the means by which the court is informed of an alleged offence. It will ultimately be the charge which the accused will be obliged to answer at any subsequent trial. An information is therefore required to describe the offence in ordinary language. Rule 100(1) of the Magistrates Courts Rules 1981 sets the detail of what an information must contain. It requires that every information, summons or other document laid or issued for the purposes of any proceedings before a magistrates court is sufficient if it describes the specific offence with which the accused is charged, in ordinary language, avoiding as far as possible the use of technical terms and without necessarily stating all the elements of the offence. It must also give such particulars as is necessary for giving reasonable information on the nature of the charge.

It is unlikely that it will be the Environmental Health Officer who would lay the information. This would most probably be done by the solicitor acting for the local authority. There is however no prohibition against the Environmental Health Officer laying the information. The Magistrates Court Act 1980 provides that any person authorised by the prosecutor may lay an information before a justice of the peace. The information must be laid before a justice or a clerk to the justice. Section 1 of the Magistrates Court Act 1980 stipulates that, upon an information being laid before a justice that any person has, or is suspected of having committed an offence, the justice may issue a summons requiring that person to appear before a magistrates court to answer the information or issue a warrant to arrest that person and bring him before the court. An information may be laid either orally or in writing. If it is given orally it does not need to be on oath unless a warrant is sought (s.1(3) Magistrates Court Act 1980). As a matter of practice all informations laid in connection with an environmental health offence will be in writing. The information is accepted as being laid when it is received at the office of the clerk to the justices; it is not necessary that the matter be put before the clerk of the justices personally. Any Environmental Health Officer laying an information would be well advised to record the date and time that the information was handed to the office. There certainly have been cases where this has proved vital in establishing that the subsequent proceedings had not become time barred under s.127 of the Magistrates Court Act 1980.

## 9.3 TIME PERIOD FOR PROCEEDINGS

There is no time limit for proceedings for an indictable offence, however s.127(1) of the Magistrates Courts Act 1980 requires that, in the absence of an express statutory provision to the contrary, proceedings for summary matters must be commenced, that is the information must be laid, within six months of the offence having been committed. For a summary offence this six-month period runs to the date the information is laid, not to the date when the accused actually appears in court (*Kennet Justices ex parte Humphrey* [1993]). Therefore in general, where an offence is triable only summarily, this is the relevant period and is an important limitation. The inter-departmental inertia which can arise when a prosecution file is passed from one local authority department to another may result in the unwary participants falling into the trap of attempting to prosecute on time expired matters. This, when viewed from the outside, is unforgivably unprofessional. The Environmental Health Officer is therefore key here; he must ensure that matters do not run out of time. In calculating the relevant time the date on which the offence was committed

## INFORMATION

**DATE:**

**THE ACCUSED:**..............................................................................

**OFFENCE ALLEGED:**

> that the accused without reasonable excuse failed to comply with a Noise Abatement Notice issued by Anytown City Council on the 13th day of September 1995 pursuant to Section 80 of the Environmental Protection Act 1990 and served on that date, in respect of the premises .............., in that the accused as the occupier of the premises failed to abate noise amounting to a nuisance caused by amplified music at .............. contrary to Section 80(4) of the Environmental Protection Act 1990

**THE INFORMATION OF**

..............................................................................

..............................................................................

Signature..............................................................

Taken before me

Justice of the Peace/Justices' Clerk

Date..............................................

**Form 9.1** Information

## INFORMATION

**DATE:** 1st August 1996

**THE ACCUSED:**

> PETER PAUL PICKLOCK
> 1 NICE STREET,
> ANYTOWN, AN12 2TN

**OFFENCE ALLEGED:**

> On the 1st day of April 1996, contrary to section 14(1) of the Food Safety Act 1990, did at the Cool Cafe, Bun Street, Anytown sell to the prejudice of the purchaser, John Smith, food, namely a pork pie, which was not of the substance demanded in that it contained an earthworm.

**THE INFORMATION OF**

> ..................................................................................
> ..................................................................................
> Signature................................................................

Taken before me


Justice of the Peace/Justices' Clerk


Date................................................

**Form 9.2** Information

is to be excluded (*Marren v Dawson Bentley & Co.* [1961]). Thus for an offence committed on the 1 January the last date for laying an information is 1 July. Were the offence to be committed on the 31 March the relevant date would be the 30 September. For continuing offences the time runs from each day the offence is committed. It is important to note that this six-month time limit does not apply to an indictable offence and therefore not to summary proceedings where the offence is triable either way. In certain instances the six-month time limit can be extended. For example section 34 of the Health and Safety at Work etc. Act 1974 has modified this rule and allows for an extension of the time period in specified cases, e.g. where a coroner's inquest is held touching the death of any person whose death may have been caused by an accident while he was at work, or a public enquiry into any death that may have been so caused.

## 9.4 DUPLICITY

It is essential that an information and summons does not charge more than one offence. Rule 12 of the Magistrates' Court Rules 1981 states that a magistrates court shall not proceed to the trial of an information that charges more than one offence. Such an information is bad for duplicity and the court is prohibited from convicting on the matters alleged. Formerly it was the case that if an information was duplicitous it had to be amended before the trial started. If it was later found that a summons was duplicitous while proceedings were underway it could not then be rectified by amendment (*Edwards v Jones* [1947]). Now the situation is that if the duplicitous matters are identified during the trial the court is obliged to request that the prosecutor elect which offence he is to proceed upon. Once identified other matters are struck out and the trial proceeds on the amended information.

Duplicity is indicated if an information charges more than one activity. In the case of *George v Kumar* [1982] the defendant was charged with 18 offences under regulation 25 of the Food Hygiene (General) Regulations 1970 (now repealed) and two under regulation 16(1)(a).

Regulation 25 required that the walls, floors, doors, windows, ceilings, woodwork and all other parts of the structure of every food room should be kept clean and should be kept in such good order, repair and condition as to:

a) enable them to be effectively cleaned, and
b) prevent, so far as is reasonably practicable, any risk of infestation by rats mice or insects.

The justices who heard the case initially found that the regulations did not create separate offences in relation to each requirement but that each regulation created one offence. The prosecution appealed. On appeal it was found that regulation 25 created two offences:

1. Failure to keep the structure of every food room clean.
2. Failure to keep the structure of every food room in good order, repair and condition so as to be effectively clean or so as to prevent any risk of infestation by rats, mice and insects.

On appeal Ormrod LJ noted that:

> prosecution authorities feel obliged to lay information in such a way that they take in every conceivable breach of the regulations and at the same time avoid the risk of having the information quashed for duplicity . . . in other words faced with the risk of submissions by the defence on duplicity, they adopt what in another context is called grape-shot therapy in the hope that they will hit the target somewhere or other.

Such an approach, though understandable, is frowned upon by the courts and therefore ought to be avoided by any Environmental Health Officers and their legal advisors. It is not duplicitous to set out common factual and legal matters in a preamble to the summons and then set out the various offences charged in numbered paragraphs if the offences are contrary to the same statutory provisions. In *Shah v Swallow* [1984] a shop owner was convicted under the Food Hygiene (General) Regulations 1970 of four out of five offences contained in an information. The five offences were contained in a single information. The first paragraph of the document contained allegations of fact in relation to each of the offences; and the second paragraph identified the regulations which created the offences. The House of Lords found that the document was a single document which contained five informations. The fact that the preamble contained details which were common to a number of separate allegations did not tie them so as to make the information duplicitous.

There is clearly a need for precision when framing the information on which a summons will be issued. It may not always be easy to decide the content of the information but it is most definitely a decision which ought to be made by the Environmental Health Officer in consultation with a legally qualified person. Consider for example section 14 of the Food Safety Act 1990. The section states that any person who sells to the purchaser's prejudice any food which is not of the nature or substance or of the quality demanded by the purchaser, shall be guilty of an offence. It is accepted by the courts that there is an area of common ground between the three words: nature, substance and quality (*Preston v*

*Greenclose Ltd.* (1975)). Yet it is possible, and indeed is required, to decide which of the three words is the most suitable. Any summons should not be so framed as to allege the offence of selling food 'not of the nature, substance or quality' This wording alleges three offences and is bad for uncertainty. Obviously the rule against duplicity holds good across the whole range of the Environmental Health Officer's work. In the case of *Wandsworth LBC v Sparling* (1987) the court found that if a local authority wishes to prosecute the owner of an HMO for breach of the Housing (Management of Houses in Multiple Occupation) Regulations 1962 (see now the Housing (Management of Houses in Multiple Occupation) Regulations 1990) it must say whether it is charging him with 'knowingly' contravening a particular regulation or 'without reasonable excuse' failing to comply with it.

### 9.5 SUMMONS

As indicated, s.1 of the Magistrates Court Act 1980 states that upon an information being laid before a justice of the peace that any person has or is suspected of having committed an offence, the justice may issue a summons requiring the person concerned to appear to answer the information. The summons should contain the nature of the information and the time and place at which the accused is required to appear. Though containing essentially the same information whatever the offence the format of a summons may change.

### 9.6 COMPLAINT

Part II of the Magistrates Court Act 1980 deals with the civil jurisdiction of the Magistrates Court. It indicates how it is possible to obtain a summons from the court by the making of a complaint. A complaint is a written or verbal reference to a justice or justice's clerk that a person or a corporation has committed a breach of the law which is not criminal. As for an information once the complaint is received at the office of the clerk the complaint is accepted as having been duly made. Section 52 of the Magistrates Court Act 1980 stipulates that where a complaint is made, a justice may issue a summons directed to that person requiring him to appear before a magistrates court acting for that area to answer the complaint. This is the means for commencing a civil action in a magistrates court where, according to s.52 of the Magistrates Court Act 1980, the court has jurisdiction.

It is not very common for Environmental Health Officers to be party to such proceedings but it is possible. For example s.82 of the Environmental Protection Act 1990 allows an individual aggrieved by the existence of a statutory nuisance to bring an action in a magistrates court. The statute provides that a magistrates court may act under this section on a complaint made by an aggrieved person. In fact a number of cases have found that the procedure contained in s.82 of the Environmental Protection Act 1990 though commenced by complaint is in fact criminal in nature (*Herbert v Lambeth LBC* (1991), *Botross v London Borough of Hammersmith and Fulham* [1995]).

## SUMMONS

**DATE:**

**TO THE ACCUSED:**

......................................................................................

**OFFENCE ALLEGED:**

that you without reasonable excuse failed to comply with a Noise Abatement Notice issued by Anytown City Council on the 13th day of September 1995 pursuant to Section 80 of the Environmental Protection Act 1990 and served on you on that date, in respect of the premises ................., in that you as the occupier of those premises failed to abate noise amounting to a nuisance caused by amplified music at .............. contrary to Section 80(4) of the Environmental Protection Act 1990

**THE INFORMATION OF :**

Prosecutor.........................................................

......................................................................

Signature...........................................................

YOU, THE DEFENDANT ARE HEREBY SUM-MONED to appear before the *MAGISTRATES COURT* sitting at .................... 10.00am on the........................ day of .............. 1996 to answer to the Information of which particulars are given hereto.

Justice of the Peace/Justices' Clerk

Communication
answer to the Summons
to be sent to

The Clerk to the Justices
Anytown Magistrates Court
1 Centre Place, Anytown

Form 9.3 Summons

# SUMMONS

INFORMATION has this day been laid before me by
JANE JONES
ENVIRONMENTAL HEALTH OFFICER
ANYTOWN CITY COUNCIL
TOWN HALL, ANYTOWN, AN1 1TN

who states that the accused

PETER PAUL PICKLOCK
1 NICE STREET,
ANYTOWN, AN12 2TN

Date of Birth          18th October 1959

On the 1st day of April 1996, contrary to section 14(1) of the Food Safety Act 1990, did At the cool cafe, Bun Street, Anytown sell to the prejudice of the purchaser, John Smith, food, namely a pork pie, which was not of the substance demanded in that it contained an earth worm.

Dated the 15th day of August 1996

To the above-named defendant
        YOU ARE SUMMONED TO APPEAR BEFORE THE MAGIS-TRATES COURT SITTING AT Anytown at 10.00 AM on 10 September 1996 to answer the said information.

Communication                    The Clerk to the Justices
answer to the Summons            Anytown Magistrates Court
to be sent to                    1 Centre Place
                                 Anytown

**Form 9.4** Summons

## FILE FRONT SHEET

Defendant(s)

| |
|---|
| |
| |
| |
| |

Details of Offences (include identification of relevant statutory provision)

.................................................................................................................

.................................................................................................................

.................................................................................................................

.................................................................................................................

Officer in Charge (include telephone number) ........................................

Statutory Time Limit .................................................................................

| Date of Hearing | Place of Hearing |
|---|---|
| | |
| | |
| | |

Plea entered. (If change of pleas record all details inc. dates)..................

Result (if appropriate include details of fines, costs and other orders of the court).

.................................................................................................................

.................................................................................................................

.................................................................................................................

**Form 9.5**  File front sheet

| FILE CONTENTS | | |
|---|---|---|
| **DOCUMENT** | ✓ | **(INDICATE NATURE)** |
| 1.  FRONT SHEET | | |
| 2.  CONTENT SHEET | | |
| 3.  DEFENDANT | | |
| 4.  INFORMATION | | |
| 5.  SUMMARY OF EVIDENCE | | |
| 6.  WITNESS LIST | | |
| 7.  WITNESS AVAILABILITY | | |
| 8.  STATEMENTS | | |
| 9.  EXHIBIT LIST | | |
| 10.  RECORD OF INTERVIEW | | |
| 11.  LIST OF PREVIOUS OFFENCES | | |
| 12.  LIST OF PREVIOUS CAUTIONS | | |
| 13.  COSTS DETAILS | | |
| 14.  PHOTOGRAPHS | | |
| 15.  COMPENSATION CLAIM | | |
| 16.  DOCUMENT | | |
| 17.  DOCUMENT | | |
| 18.  DOCUMENT | | |
| 19.  DOCUMENT | | |
| 20.  DOCUMENT | | |
| 21.  DOCUMENT | | |
| 22.  DOCUMENT | | |

**Form 9.6**  File contents

## DEFENDANT

**Individual**

Surname ..............................................................................................................

Forename(s) .......................................................................................................

Address ..............................................................................................................

............................................................................................................................

D.o.B ..................................................................................................................

**Company**

Name ...................................................................................................................

Corporate/Unincorporate ..................................................................................

For Partnership give names of all partners

............................................................................................................................

............................................................................................................................

Registered Office (If no registered office indicate place of business) or Principal office of the Partnership

............................................................................................................................

............................................................................................................................

**Offence Details**

Date .............................. Location ..................................................

Nature (Statutory Provision and Description)

............................................................................................................................

............................................................................................................................

............................................................................................................................

............................................................................................................................

Others Involved

............................................................................................................................

............................................................................................................................

**Officer in Charge** ..........................................................................................

**Form 9.7** Defendant

## CASE SUMMARY

NAME ............................ SIGNED ............................ DATE ............

**Form 9.8** Case summary

# Court Proceedings　　**10**

## 10.1 CONDUCTING PROCEEDINGS

Section 1 of the Prosecution of Offences Act 1985 created the Crown Prosecution Service (CPS). The CPS is a national prosecution agency for England and Wales headed by the Director of Public Prosecutions. The service is organised on an area basis, there being 13 areas and over 100 branch offices, and is responsible for the conduct of all proceedings instituted on behalf of a police force (s.3(2)). No such similar agency exists for prosecutions brought by other public agencies. The 1985 Act, at s.6, did however preserve the right for private individuals and certain statutory bodies to commence criminal proceedings. So it is that the Health and Safety Executive, the Inland Revenue and local authorities are all able to prosecute on relevant criminal matters independently of the Crown Prosecution Service and each other. Whilst the former bodies have the benefit of a national framework, local authorities are very much their own masters when it comes to decisions relating to prosecution and are not usually considered to be major participants in the criminal justice system. Nevertheless there are many occasions when a local authority will find itself a party to criminal proceedings. These will most commonly arise when an authority is undertaking one of its many enforcement activities under the numerous statutory provisions it enforces. The general power of a local authority to institute proceedings is to be found in s.222 of the Local Government Act 1972, which states that where a local authority considers it expedient for the promotion or protection of the interests of the inhabitants of their area:

(a) they may prosecute or defend or appear in any legal proceedings and in the case of civil proceedings may institute them in their own name, and

(b) they may, in their own name, make representations in the interests of the inhabitants at any public inquiry held by or on behalf of any minister or public body under any enactment.

The Environmental Health Officer will under these provisions thus find himself engaged in any number of prosecutions undertaken by a local authority. Such involvement is usually limited to being a prosecution witness but this need not be the only role. Section 233 of the Local Government Act 1972 gives the means by which an Environmental Health Officer may personally present a case in a magistrates court. This section provides that any member or officer of a local authority who is authorised by that authority to prosecute or defend on their behalf or to appear on their behalf in proceedings before a magistrates court shall be entitled to prosecute or defend or to appear in any such proceedings and, notwithstanding anything contained in the Solicitors Acts 1957 to 1965, to conduct any such proceedings although he is not a solicitor holding a current practising certificate. Though rare, there are few practical reasons why EHOs should not present their own cases. Certainly some legal practitioners have expressed a view that some Environmental Health Officers, with training and a desire to appear before the courts, would be more than competent to undertake this task. Matters of procedure and evidence could be adequately addressed by further training and such an option may allow the more technical elements of a case to be more easily and perhaps more vigorously presented to the courts. It may even be that such an approach would offer a greater degree of control over the prosecution of cases and with it secure a greater degree of uniformity and consistency, both when deciding to prosecute and in any subsequent prosecution.

## 10.2 CRIMINAL PROCEDURE

For criminal matters in this country there are two modes of trial:

(a) Trial on Indictment
(b) Summary Trial.

The former takes place in the Crown Court before a judge and jury; the latter takes place in a Magistrates Court either before a lay bench or a stipendiary magistrate. Where the matter will be heard will depend on the type of offence. Criminal offences are classified broadly according to seriousness into three categories: Summary offences, Indictable offences and Offences triable either way. Each is defined in Schedule 1 of the Interpretation Act 1978 (see Chapter 1) in the following way.

### 10.2.1 Summary Offences

These are offences which if committed by an adult are triable only summarily. Summary offences, which can normally only be dealt with in a

Magistrates Court, are the least serious criminal offences and make up the bulk of matters put before the Magistrates Courts.

### 10.2.2 Indictable Offences

These are offences which if committed by an adult are triable on indictment, whether exclusively so triable or triable either way. These are the most serious type of criminal offence and the Magistrates Court has no jurisdiction to try them; they must go to the Crown Court. Magistrates Courts will deal with such cases only to address issues of bail and committal.

### 10.2.3 Offences Triable Either Way

These are offences which if committed by an adult are triable either on indictment or summarily. The decision where the matter is tried is one for both the magistrates and the accused.

Though the national mode of trial guidelines for magistrates[1] do not address themselves expressly to the Environmental Health offender, guidance contained in s.19 of the Magistrates Courts Act 1980 is nevertheless relevant and requires magistrates to have regard to the following matters when deciding whether an offence is suitable for hearing in the Magistrates or Crown Court:

1. The nature of the case
2. Whether the circumstances make the offence one of a serious character
3. Whether the punishment which a magistrates court would have power to inflict for it would be adequate
4. Any other circumstances which appear to the court to make it more suitable for the offence to be tried in one way rather than the other
5. Any representations made by the prosecution or the defence.

These guidelines and experience will suggest that by far the most common court for the Environmental Health Officer to appear in is the Magistrates Court and it is to this forum that the following is addressed.

### 10.3 SUMMARY TRIAL

#### 10.3.1 MAGISTRATES COURTS

Magistrates Courts are the most junior or inferior of the criminal courts and are probably the one the Environmental Health Officer is most likely to encounter professionally. Primarily a court of criminal jurisdiction it is not exclusively so and does retain some civil and administrative functions, for example matters relating to licensing and cases in the Family Proceedings court.

### 10.3.1.1 Magistrates

The emphasis placed on the Magistracy is a distinctive feature of the English legal system. Indeed it is in the Magistrates Courts that the vast majority of criminal cases are dealt with. There are in England and Wales some 30 000 magistrates.[2] They are either appointed by the Lord Chancellor or, in the Duchy of Lancaster, by the Chancellor of the Duchy.[3] The vast majority of these are 'lay Magistrates', who unlike their stipendiary colleagues are not required to be qualified lawyers. They are simply required to be 'persons of good character and repute who are able to command the confidence of both the public and their colleagues and possessing the capability to act judicially and thus to make fair and proper decisions'. They are required, and are trained, to recognise personal prejudices and to set them aside. They must understand what facts are relevant to a case and be able to think clearly and logically. Generally Magistrates are, like the professional judiciary, expected to display qualities of fairness, judicial temperament and a willingness to hear both sides. However, unlike the professional judiciary, they are also required to be seen to be (and to be) reasonably representative of the community to which they belong.[4]

### 10.3.2 OFFENCES

There is at the time of writing a desire in certain quarters to limit the right of an accused to elect for jury trial as this is seen to be a very expensive procedure, especially when dealing with a less serious either way offence. The advantage of a summary trial is that it is relatively quick and therefore inexpensive when compared to the more time consuming and costly approach of the Crown Court. At a time when the legal aid budget is coming under scrutiny it is not difficult to see why the limitation on rights to elect jury trial is a topical issue.

All of the offences encountered by the Environmental Health Officer will be indicated as triable summarily or either way. For example under the Environmental Protection Act 1990 section 23(1)(2) provides that the carrying on of a prescribed process without an authorisation or in breach of a condition of an authorisation, is an offence and the maximum penalty is, on conviction in the Magistrates Court, a fine of £20 000 and on conviction by the Crown Court an unlimited fine and/or two years' imprisonment; clearly this is an either way offence. The same Act, as amended by the Environment Act 1995, stipulates that obstructing an authorised person in the exercise of their powers of investigation carries a maximum penalty of a fine of £2000 on conviction by the magistrates court; this is a summary only offence.

Sections 7, 8 and 14 of the Food Safety Act 1990 are triable either way having penalties on indictment of up to two years prison, an unlimited fine or both and on summons six months, a £20 000 fine or both. Offences contrary to section 33(1)(a) and (b) of the Food Safety Act 1990 are triable summarily only with a maximum penalty of £5 000, three months or both.

The very fact that a number of offences are either way offences imposes on an Environmental Health Officer an obligation to have an opinion, informed by the findings of his investigation, on the appropriate forum for trial.

### 10.4  BEFORE THE HEARING

Ambushing of a defendant by the prosecution has no part to play in criminal procedure. So it is that for either way offences, but not summary only offences, the Magistrates Courts (Advance Information) Rules 1985 impose an obligation on the prosecution to provide the defence with details or a summary of the case they intend to present. Such advance information must contain either a copy of the written statements and supporting documents on which the prosecution propose to rely or a summary of the facts and matters of which the prosecution propose to adduce evidence.

As soon as practicable after a person has been charged with an offence in proceedings in respect of which the Rules apply or a summons has been served on a person in connection with such an offence, the prosecutor is obliged to provide him with a notice in writing explaining the effect of Rule 4 of the Rules and setting out the address at which a request under the Rule may be made. Rule 4 states that if, in any proceedings in respect of which these Rules apply, before the magistrates court considers whether the offence appears to be more suitable for summary trial or trial on indictment, the accused or a person representing the accused requests the prosecutor to furnish him with advance information, the prosecutor must, subject to what is contained in Rule 5, furnish him as soon as practicable with either:

(a)  a copy of those parts of every written statement which contain information as to the facts and matters of which the prosecutor proposes to adduce evidence in the proceedings, or

(b)  a summary of the facts and matters of which the prosecutor proposes to adduce evidence in the proceedings.

Rule 5 limits the extent of the obligation to disclose. If the prosecutor is of the opinion that the disclosure of any particular fact or matter might lead to any person on whose evidence he proposes to rely in the proceedings being intimidated, he shall not be obliged to comply with those requirements in relation to that fact or matter. If a prosecutor considers that he is not obliged to comply with the requirements of disclosure imposed by the Rules, he must give notice in writing to the person who made the request to the effect that certain advance information is being withheld.

Environmental Health professionals may on occasion view this requirement as 'showing your hand' to the opposition. It is however a basic proposition of law, and one which should be respected, that any individual accused of an offence should have details regarding the accusation in order that he may defend himself against the allegation. Disclosure offers the opportunity for this. It is no part of the Environmental Health Officer's duties to spring surprises on an unwary defendant. Also disclosure allows the defence solicitor to take full instructions from his client on all relevant matters. This is beneficial for all parties as it will minimise the likelihood of delay attributable to unexpected adjournments. An overwhelmingly strong case, when disclosed, may result in an early guilty plea, thereby avoiding the need for a costly trial and offering benefits to both sides.

Though the requirement for disclosure relates only to either way offences it is not unusual for disclosure to be requested and to occur where the offence is triable only summarily.

## 10.5 THE HEARING

### 10.5.1 PRELIMINARIES

Bringing matters before the courts is time consuming and the processes involved are often hidden from many of the participants. It is therefore vital that the Environmental Health Officer should as a routine part of his duties take the time to explain the court procedure to all his witnesses. For an Environmental Health Officer the giving of evidence may never be pleasant but it is possible by drawing on experience and good preparation to minimise the unpleasantness. For someone who is not a professional witness the experience may be much more worrying. It is all too easy when taken up with the events of a trial to forget just how uncertain, how confused, how frightened a witness may be. As this anxiety could affect the quality of the evidence they may give, a few careful and informative words from the Environmental Health Officer as to what they might expect will go a long way to making matters less fraught for all parties. This should not mean the Environmental Health

Officer coaching his witnesses, but a few moments spent reassuring nervous witnesses, telling them about the personnel of the court, the sequence of events, what they will be expected to do and what they are entitled to expect from the court is little more than basic courtesy and may pay dividends when they come to be called. Once in court the Environmental Health Officer may look to the lawyers to ease the nerves of a timid witness, but before that point it is the Environmental Health Officer who is likely to have been in regular contact with them and thus best able to inspire confidence.

### 10.5.2 IDENTIFICATION

Before proceeding to hear the case it is important that the person before the court is formally identified and the matters against them are put. It is the clerk who will normally identify the person accused of the offence and an exchange similar to that following will occur.

**Usher:** *Case number six your worships, Grace Green.*
**Clerk:** *Mr Black for the Local Authority and Ms Green represented by Ms White your worships.*
*(to the accused) Is your full name Grace Green?*
**Green:** *Yes.*
**Clerk:** *Date of birth 18th March 1965?*
**Green:** *Yes*
**Clerk:** *Home address 1 Railway Terrace, Anytown ?*
**Green:** *Yes.*
**Clerk:** *Ms Green it is alleged that on the 9th October 1995 in Anytown you, without reasonable excuse, failed to comply with the requirements of an abatement notice served on you by the Anytown City Council, under the provisions of Part 3 of the Environmental Protection Act 1990. Do you understand this matter?*
**Green:** *Yes.*
**Clerk:** *Ms Green to that matter do you plead guilty or not guilty?*

Having correctly identified the defendant the issue of their plea to the charge will then need to be considered.

### 10.5.3 THE PLEA – GUILTY

A hearing before a court may be to decide the question of guilt or innocence or in the event of a guilty plea to decide sentence. A guilty plea occurs when an accused person freely and clearly admits the offence in clear and unambiguous terms. Usually this will be after the matter has been put to him and after confirmation of his understanding of the matter has been obtained. In the case of a guilty plea for an either way offence the sequence followed is:

1. The nature of the allegation is read out. Section 19(2)(a) of the Magistrates Court Act 1980 requires that before considering the question of mode of trial the charge must be read to the accused.
2. If the offence is an either way offence the defendant will then have an explanation of the right to elect for jury trial made to him.
3. There will then be a brief outline of the case for the prosecution and representations made as to the most suitable method of trial. Section 19(2)(b) stipulates that first the prosecutor and then the accused is given an opportunity to make representations as to which mode of trial would be more suitable. The prosecution might at this stage suggest that the magistrates court has insufficient powers of sentence to deal with the matters before them. The defence may contend they have sufficient powers.
4. The magistrates will then decide if they have sufficient powers to deal with the case. Section 19(3) of the Magistrates Court Act 1980 details the matters which the court is to consider when deciding on a venue for the trial.
5. Section 19 of the Magistrates Courts Act 1980 requires magistrates to have regard to the following matters in deciding whether an offence is more suitable for summary trial or trial on indictment:
   - the nature of the case
   - whether the circumstances make the offence one of a serious character
   - whether the punishment which a magistrates court would have power to inflict for it would be adequate
   - any other circumstances which appear to the court to make it more suitable for the offence to be tried in one way rather than the other
   - any representations made by the prosecution or the defence.

To assist the magistrates in reaching a decision, National Mode of Trial Guidelines exist and detail certain relevant matters when making a decision. These are:

(a) the court should never make its decision on the grounds of convenience or expedition
(b) the court should assume for the purpose of deciding mode of trial that the prosecution version of the facts is correct
(c) the defendant's antecedents and personal mitigating circumstances are irrelevant for the purpose of deciding mode of trial

(d) the fact that the offences are alleged to be specimens is a relevant consideration; the fact that the defendant will be asking for other offences to be taken into consideration, if convicted, is not

(e) where cases involve complex questions of fact or difficult questions of law, the court should consider committal for trial

(f) where two or more defendants are jointly charged with an offence and the court decides that the offence is more suitable for summary trial, if one defendant elects trial on indictment the court may not proceed to deal with all the defendants as examining justices in respect of that offence. A young person jointly charged with someone aged 18 or over should only be committed for trial if it is necessary in the interest of justice

(g) in general, except where otherwise stated, either way offences should be tried summarily unless the court considers that the particular case has one or more of the features set out elsewhere in the guide *and* that its sentencing powers are insufficient.

6. If the magistrates decide that the case is too serious and that they would not have adequate powers to deal with the matter they may decline jurisdiction and commit the case to the Crown Court. The accused is not consulted on this and has no right to object.

7. If the magistrates court accepts the case the accused is then given the option to have the case heard before the Magistrates or elect for trial by judge and jury at the Crown Court. Before he makes his decision he is warned that should he elect for trial in the Magistrates Court and that court subsequently, after conviction, decides that it does not have adequate powers of sentence it may commit the matter to the Crown Court for sentence (s.38 Magistrates Court Act 1980).

8. If the accused elects to stay in the Magistrates Court matters may proceed there.

9. If he elects for jury trial then matters will removed to that court.

10. If he remains in the Magistrates Court, the accused will then enter his guilty plea.

11. Any matters which are to be taken into consideration (TICs) are handed in and formally admitted at this stage.

12. The prosecution will then read out the facts of the case and make any claim for costs and compensation. Costs must be reasonable and are at the court's discretion.

13. The defence will then speak in mitigation.

14. The Magistrates will then impose their sentence and make any order for costs.

15. If after hearing the case the court does not feel it has adequate powers of sentence it may commit the case to the Crown Court for sentence.

### 10.5.4 THE PLEA – NOT GUILTY

In the event of a not guilty plea a trial will have to be fixed at which the evidence will be heard. Statements will then normally be served by the prosecution on the defence.

#### 10.5.4.1 Pre-Trial Review

There will usually be a pre-trial review to ascertain which witnesses will be required and how long the case will be likely to take. This will not usually involve the Environmental Health Officer and is largely an administrative matter between the respective lawyer and the courts.

## 10.6 THE TRIAL

### 10.6.1 WITNESSES

It is a rule of practice that witnesses should remain outside court until called to give their evidence. Therefore anyone present in the court for the purposes of giving evidence should at the point the trial commences, if they have not already done so, leave the courtroom. It would be at this stage that the Environmental Health Officer would be expected to leave the courtroom and take a seat outside. All courts have seating just outside and it is probably best for the Environmental Health Officer to sit with or close to any of his witnesses. They will, naturally, look to him for clarification of anything they don't understand and will find his presence reassuring. The Environmental Health Officer must, for his part, seek to ensure that the normal tension that all parties will be experiencing does not turn into a more damaging paralysis. The professionalism of the officer outside court can be just as significant as that displayed inside.

### 10.6.2 OPENING REMARKS

Rule 13 of the Magistrates' Courts Rules 1981 deals with the order of evidence and speeches in a summary trial. Where there is to be a trial of an information and the accused is not pleading guilty the prosecutor will call the evidence for the prosecution and before doing so he may address the court (r.13(1)). There is no obligation on a prosecutor to give an opening speech, though the opportunity is invariably taken. Such opening remarks are meant to assist the court in understanding the nature of the case and to indicate the main issues which will arise in evidence. It is very useful for the court to have a brief outline of the case since it assists it in understanding the significance of the evidence it will hear later.

Typically a prosecutor will stand and open his case in the following way:

*Prosecuting Solicitor (stands and addresses the bench): Your worships, the principal prosecution witnesses in this case are Mr Blue and Mr Scarlett from the local authority. Mr Blue will testify that at approximately 1.00 am on the morning of the 9th October 1996 he, accompanied by another officer, Mr Scarlett, was on duty outside Ms Green's house. The officers were engaged in the monitoring of noise coming from 1 Railway Terrace, Anytown. This is Ms Green's house. Mr Blue monitored the noise using the most up to date equipment. The noise consisted of amplified music and whistles. The officer will state that after a reasonable period of monitoring he was satisfied that a contravention of a previously served noise abatement notice was occurring. The notice was originally served in October 1995 as a result of numerous complaints received from neighbours concerning excessive noise from earlier parties held by Ms Green. Your Worships, the Environmental Protection Act 1990 empowers a local authority, if satisfied that a noise nuisance has occurred, is occurring or, as here, is likely to recur, to serve a legally enforceable notice prohibiting that recurrence. Any person aggrieved by such a notice is, within a limited period, entitled to appeal against the notice. No such appeal was ever lodged by Ms Green. You will hear that despite having received such a notice Ms Green nevertheless ignored its requirements and held yet another noisy party. This was without reasonable excuse and was in contravention of the said notice. She thus committed an offence.*

Prosecution witnesses will then be called in an order at the prosecutor's discretion. It can assist the court to understand the sequence of events if witnesses are called in broad chronological order. Events unfold in a particular sequence and are usually best retold in that sequence.

The court will normally hear oral evidence but for certain evidence written statements are acceptable and may be read before the court. Primarily these are for statements under section 9 of the Criminal Justice Act 1967 when evidence is undisputed. The use of these statements saves both witness and court time. Witnesses are not required to attend and unnecessary examinations can be avoided.

10.6.3 EXAMINATION IN CHIEF

Examination in chief is the name given to the initial questioning of a witness by the side that has called him. e.g. prosecution witness examined by the prosecution. The object in examination in chief is to obtain, by putting proper questions, which are not in leading form, relevant and admissible evidence which supports contention of the party who calls the witness.

*Archbold*

In this the examination of the witness should be relevant and must relate to the facts in issue or those which are relevant to the issue. Leading questions are not generally permissible in examination in chief nor in re-examination and may not be put if the other party objects. Such questions may however be correctly asked in cross-examination. A leading question is simply one which suggests its own answer. For example were an Environmental Health Officer to be asked: 'Did you see any dirt?' 'Was the guard loose?' these would be leading questions as each is put so as to suggest the answer that is wished: 'Yes there was dirt'; 'Yes the guard was loose.'

In practice a prosecutor may occasionally be permitted to lead the witness. This will occur where there are matters of fact which are not in dispute. Here one side may lead after first obtaining the consent of the other. This serves to speed up the procedure, thereby saving court time and money, yet without any undue prejudice to an accused. Thus an Environmental Health Officer may be asked: 'Did you go to 1 Railway Terrace?'; 'Did you use a sound level meter?'

Though leading questions, these matters of fact may not be in dispute. To allow the use of these leading questions will assist the court at no penalty to an accused. It is of course possible for the permission to lead to be terminated by the opposition; at that point non-leading questions must once again be put.

Examination in chief is the process by which one side, here the prosecution, sets out the evidence in support of their case. The opportunity of the other side to challenge what is said occurs after examination in chief, under cross-examination.

## 10.6.4 CROSS-EXAMINATION

The object of cross-examination is to test the truth and accuracy of the matter being asserted by the witness. It may have as its aim a desire to show a witness as unreliable, biased, uncertain or simply not to be believed. A witness may be discredited and at the very least cross-examination offers an opportunity to display to a court that the matters revealed in examination in chief are not accepted unchallenged, nor in all their particulars. This part of the proceedings has broad objectives. The cross-examiner must put his case to the witness. That is, he must question on those points on which accounts diverge. Cross-examination may elicit extra, and to the defence, useful facts, and it may discredit the evidence previously given in examination in chief.

One of the most obvious ways to discredit evidence is to discredit the witness. To avoid this the Environmental Health Officer should realise that he may be challenged on his veracity, the accuracy of his record of events, his integrity, his reasons for recollection, his experience and knowledge. He should anticipate questioning which will seek to expose discrepancies, omissions, errors, exaggeration etc. in what he has said and he should know that the person undertaking the cross-examination will be particularly looking for matters where witnesses do not agree. Unlike an examination in chief leading questions are permissible in cross-examination. These may exceptionally be put such as: 'This is not true is it?'

More usually they will be prefaced with the line much favoured by dramatists: 'I put it to you.'

This can prove to be a testing time for any officer. The cross-examination may be bullying and aggressive or it may be slow and steady; it will always be firm. Preparation for this can be difficult but some matters can be anticipated. Before the hearing an Environmental Health Officer must speak to his solicitor and consider questions likely to arise under cross-examination. By anticipating such questions the best response can be decided upon. The answer must in no way mislead and on no account should an officer lie. But questions on the quality of observation, recording and interpretation may be anticipated and unnecessary problems avoided. In like manner rehearsal with fellow officers can help to identify any technical, as contrasted to legal, difficulties which may arise. For example, the technical issue of the calibration of a sound level meter will have bearing on the legal issue of whether the levels measured contravened the requirements of a statutory notice.

As a general rule any witness who modestly and carefully gives answers to the questions asked, does not get angry, does not become abusive or insolent, but sticks to his story can withstand a cross-examination and indeed such a witness can even result in the cross-examination assisting the prosecution's case. For here a cross-examiner may have merely achieved a repetition of the examination in chief.

On cross-examination it is permissible to inspect any document used for memory refreshing in examination in chief. It is here that a failure to observe the disciplined keeping of notebooks, inspection records etc. could prove to be embarrassing.

### 10.6.5 RE-EXAMINATION

Re-examination follows cross-examination. Any re-examination must be limited to, and arise from, the cross-examination and cannot introduce fresh evidence. Once again leading questions are not permitted to be put to the witness.

## 10.6.6 NO CASE TO ANSWER

It is important to remember that it is for the prosecution to prove their case. There is no obligation on the defence to establish innocence. Generally speaking, the prosecution case is as strong as it is going to get at the point when it has presented its case but before the defence responds. It is open to the defence at this point to submit to the court that the prosecution case as presented discloses no case to answer. Usually this will be because the prosecution has failed to submit evidence on some essential point in the case or that the evidence submitted is unreliable and it is not safe to convict on this standard of evidence. Clearly in an ideal world this ought not to be possible. It is hard to resist the argument that any case which could be dismissed at this point should have not been put before the courts in the first place. Nevertheless it can and does happen.

The High Court has given guidance to the court on the submission of no case to answer in a 1962 Practice Direction on the responsibilities of justices in cases of submission of no case. Lord Parker CJ said that as a matter of practice justices should be guided by the following considerations.

A submission that there is no case to answer may properly be made and upheld:

(a) when there has been no evidence to prove an essential element in an alleged offence;

(b) when the evidence adduced by the prosecution has been so discredited as a result of cross-examination or is so manifestly unreliable that no reasonable tribunal could safely convict on it.

Apart from these two situations a tribunal should not in general be called on to reach a decision as to conviction or acquittal until the whole of the evidence which either side wishes to tender has been placed before it. If, however, a submission is made that there is no case to answer, the decision should depend not so much on whether the adjudicating tribunal (if compelled to do so) would at that stage convict or acquit but on when the evidence is such that a reasonable tribunal might convict. If a reasonable tribunal might convict on the evidence so far laid before it, there is a case to answer.

Assuming a case is found the defence will then present its case and call any evidence and be examined as before.

## 10.6.7 DEFENCE

The case for the defence proceeds in an order similar to the prosecution and is governed by the Magistrates Courts Act 1980 and Police and

Criminal Evidence Act 1984. If the accused wishes to give evidence then the time for taking this evidence is governed by s.79 of the Police and Criminal Evidence Act 1984 . This states that if at the trial of any person for an offence

(a) the defence intends to call two or more witnesses to the facts of the case; and
(b) those witnesses include the accused, then the accused shall be called before the other witness or witnesses unless the court in its discretion otherwise directs.

At the conclusion of the evidence for the defence the prosecutor may call evidence to rebut that evidence (rule 13(4) Magistrates' Court Rules 1981).

### 10.6.8 CLOSING SPEECH

The defence may, but usually will not, make a speech on opening. To do so would deprive them of the advantageous opportunity to make a closing speech at the end of all the evidence. If the court permits the prosecution to make a closing speech this must be made before the defence speaks as the defence is always the last to be heard (rule 13(6) Magistrates' Court Rules 1981).

### 10.7 GIVING EVIDENCE

### 10.7.1 PREPARATION

It is often in the giving of evidence that months of hard work will culminate. Needless to say it is vital that as much effort and attention to detail is invested in this stage as in any of the earlier stages.

The process of giving evidence in a court of law can be amongst one of the most nerve-racking experiences an Environmental Health Officer may have to undergo. It is impossible to stop anyone so inclined from becoming nervous, but the Environmental Health Officer must realise how important it is that he controls these nerves so that they do not become the means by which a case, which should have been won, is lost. Once again preparation is the key. The officer must ensure that all matters under his control are indeed controlled. There will be quite enough over which the Environmental Health Officer will not have control. It is thus much better if those things which can be regulated are. It is possible to look at what can be done before and during the giving of evidence to, if not remove nerves, go some way to ensuring that they do not get entirely out of hand.

10.7.2 BEFORE THE HEARING

Firstly it should be the aim of the officer to arrive early at the court. This serves several purposes. It gives that extra margin should traffic or some other matter conspire to make the officer late; it also ensures that he will have time to settle to the atmosphere of the court and make his presence known to his solicitor and witnesses before being called. An officer should therefore use his initiative and find the relevant court. There may be court lists posted up stating which court the case is to be heard in. In any event it is a good idea for the officer concerned to make his presence known to the Usher. In the Magistrates Court the usher will be the only person wearing a gown, and therefore ought to be easy to find. In the Crown Court though gowns are worn by more than the usher, ushers do not wear wigs and are not easily mistaken for barristers. It is the usher who will eventually call the officer from the waiting area into the court. Once the officer has established this contact he should remain in the building and certainly not leave without consent. He cannot know when he is to be called and must remain close by so as not to delay the proceedings of the court. It is a good idea not to stray too far from the court room.

As a professional the need to be dressed appropriately is self evident. For men this means a suit and for women an equivalent 'sober' method of dress. Officers are not there on a fashion show, their purpose is to give their evidence calmly and carefully. Therefore shows of ostentation are to be avoided, overly bright ties are probably not the best idea, nor would it be a good idea to go in 'dripping' in gold jewellery. For men do try to remember to remove pens from the breast pocket of a jacket. Officers who have taken time and trouble over their appearance may appear more credible when claiming that they have taken time and trouble over their investigation. In short the officer should look the part. The officer must ensure that he has all necessary papers, particularly all authorisation documentation, with him. It is much too late and far too disconcerting to discover as one is about to enter court that a key document has been left back at the office or that if challenged the officer cannot prove that he was authorised to investigate matters. The officer should carry all identification and authorisation documents and know which pocket he has them in.

10.7.3 WHEN IN THE BOX

**10.7.3.1 Swearing**

The officer should enter the witness box and if swearing raise the testament in either hand. The swearing of witnesses is regulated in a Magistrates Court by s.98 of the Magistrates Court Act 1980, which requires that all evidence before the court shall be on oath. The form of the oath is prescribed by the Oaths Act 1978 (1978 c.19). Section 1(1) of the Act stipulates that any oath may be taken in the following form and manner. The person taking the oath must hold the New Testament, or if of the Jewish religion the Old Testament, in his uplifted hand and say: 'I swear by Almighty God, [these are the only prescribed words] that the evidence I shall give shall be the truth the whole truth and nothing but the truth.'

For someone who is neither Christian nor Jew the oath may be administered in any lawful manner, for example a Muslim may use the Koran, a Hindu the Gita or a Sikh the Adi Granth. Courts can generally cope with all religions and in any event an individual for whom a holy book or its equivalent is not available, may always affirm. Unlike in popular fiction there is no need to say 'so help me God', nor is there, as is the case in Scotland, any need to swear with the free hand uplifted. Though should anyone wish to employ the latter technique they would not be stopped. The Scottish form of oath is dealt with by s.3 of the Oaths Act 1978. Where an oath has been duly administered and taken, the fact that the person who took it had at the time of taking no religious belief has no effect on the validity of the oath. Section 5 of the Oaths Act 1978 deals with those who do not wish to swear and provides that any person who objects to being sworn shall be permitted to make a solemn affirmation instead of taking an oath. Section 6 stipulates the form for an affirmation and it should be remembered that an affirmation is of the same force and effect as an oath. Such an affirmation would be given in the following way.

'I, John Smith, do solemnly and sincerely and truly declare and affirm that the evidence I shall give shall be the truth the whole truth and nothing but the truth.'

One other oath which may be professionally encountered by an Environmental Health Officer is that of an interpreter. This would be administered broadly as follows. I swear by almighty God that I will well and truly [sometimes faithfully] interpret the … language into the English language and the English language into the … language and true explanation make to the court and to the witness and to the defendant as shall be required.'

### 10.7.3.2  Giving Evidence

After swearing the officer may adopt the approach of the police and identify himself by name, designation and employing authority. If an officer proposes to do this it is wise to first agree this with his solicitor. To do otherwise may give rise to the situation where the first actions of both the officer and his solicitor are to speak over each other; not the best start.

If in a magistrates court speak to the magistrates when giving answers to questions. In practice it is to the Bench Chairman that you speak. The acceptable form of address if speaking to a single magistrate (usually the bench chairman) is Sir or Madam. The response 'Your Worships' should be retained for those circumstances when an officer finds himself addressing the bench as a whole. In the Crown Court the judge, if a High Court judge, is addressed as 'My Lord', if a Recorder or a Circuit Judge he or she is addressed as 'Your Honour'. It is part of the choreography of the court that, though an advocate may ask the question, it is to the bench, or if in the Crown Court to the judge and jury, that the witness should direct his response. As a professional witness the Environmental Health Officer must be aware of this. In practice it is common to find that the witness box is so placed that a witness may face the questioner while being questioned but, with little effort, turn to respond to the bench.

The professional witness must speak clearly and sufficiently loudly that the court can hear. When giving evidence the Environmental Health Officer should not look down nor mumble into his notebook, nor should he speak so quickly that the court finds it difficult to follow his evidence. To avoid this it is a good idea to 'follow the pen'. What this means is that the professional witness should be guided by watching the pen of the bench chairman or the judge. The chairman will certainly be taking a note of the officer's evidence, therefore the officer should watch the pen of the bench chairman and take his cue from this as to the speed at which he should speak. A professional witness should not fidget, drum his fingers etc. If an officer is unable to control such actions he should grip the sides of the box. This is not an unusual stance to see a witness adopt and will not attract comment; it will, however, help to control those distracting activities of the hand which are brought on by nerves. The courtesy shown to the bench should be extended to all in court even, perhaps especially, to the advocate for the other side. It is important that an Environmental Health Officer does not lose his temper or display aggression under cross-examination. As a professional witness an Environmental Health Officer must give his evidence in an honest, reasonable, controlled and measured way. This is impossible if an officer has lost control of his temper and may be what the cross-examiner wanted. If the cross-examination is intense, as a professional witness the Environmental Health Officer must be prepared and able to stand his ground when dealing with the questions put, while not reacting to his examiner. The officer must not allow words to be put into his mouth. Typical examples of this type of approach would be when a question is put in the form: 'would you agree . . .' . This is a not uncommon technique to try to get you to agree with an alternative account to the one given. It attempts to close down the alternative responses available to a question and thereby aid the cause of the side putting the question. If you do not agree, say so. An acceptable response to this would be 'I could not argue with that'. Such a response whilst accepting the truth of what is said does not simply respond in the limited way anticipated. An officer should not be tempted to express an opinion when he is not qualified. For example the question; 'Was he unwell ?' should be met with a response 'I cannot say, I am not a doctor'. Whereas the question 'Did he look unwell?' can be met with a yes or no response. The Environmental Health Officer must always aim for precision in his answer. His answer should be based on fact, contain all appropriate detail and relate to what was actually seen, heard etc. He must always aim to match his answer to the question put, without elaboration or enlargement. To achieve this the Environmental Health Officer must take advantage of the Memory Refreshing Rule and seek to refresh his memory, in the witness box, from any document which he made contemporaneously with the events to which he is testifying. If a question is not heard ask for it to be repeated. If unclear the Environmental Health Officer should seek clarification. If after clarification the question is still not understood then the officer must say so. Similarly if the answer is not known this must be admitted. Any attempt by an officer to bluff his way through will be quickly detected and will be damaging to his evidence. If on reflection the Environmental Health Officer feels he has given an answer which may have given the wrong impression he must be prepared to say so. The Environmental Health Officer may be responsible for the production of exhibits, especially those linked to his statement. He must ensure that all exhibits are labelled, available and produced for the court.

It is important to recognise the seriousness of the proceedings. There is no room for jokes or flippancy when giving your evidence. If you are seeking to convince the court of the seriousness of the situation humour is not the most appropriate means.

## 10.8 THE PLEA – THE UNACCEPTABLE GUILTY PLEA

### 10.8.1 THE NEWTON HEARING

This is sometimes termed the 'The Unacceptable Guilty Plea', which is not to be confused with an equivocal guilty plea, where an accused person though admitting guilt does so in such a way as to indicate he is really pleading not guilty. For example a defendant might say: 'I didn't do it but I can't afford the time to keep coming here [to court] so I plead guilty, and let's get it over with.'

In such cases the court is under an obligation to ensure that any plea entered is not equivocal. It cannot proceed to deal with a case where the plea is ambiguous. In such cases the court will deal with the matter as if a not guilty plea was entered.

An unacceptable guilty plea will occur when an accused person having pleaded guilty then proceeds to give an incorrect account of the situation or an account of the facts so much at odds with that of the prosecution that it will be likely to affect the outcome in terms of sentence. For example a motorist may admit that he was speeding but may claim that he was only doing 10 miles per hour in excess of the limit whilst the police claim his speed was 40 miles per hour over, or the accused accepts there was hygiene problem in his kitchen but alleges it was very minor when the Environmental Health Officer alleges the premises were in a filthy condition. In these cases what is commonly called a Newton Trial or a Newton Style Hearing may occur after the case of *R v Newton* (1983). In *Newton* the accused pleaded guilty to a charge of buggery with his wife. The prosecution claimed that this had been without consent. In mitigation it was claimed that there was consent. This presented the court with as sharp a divergence of fact as could have been imagined. The Court of Appeal gave guidance as to the choices available to a court in such circumstances. These are:

- to hear evidence from both sides and come to a conclusion on the matter which is at the root of the problem
- hear no evidence but listen to submissions. But where there is a substantial clash between the two sides the court must come down on the side of the defendant.

If confronted by such a situation the Environmental Health Officer may deny the alternative account being given as correct, state that these facts are not admitted and in that situation the Environmental Health Officer can then expect to both give evidence and be cross-examined. Essentially the court is not then concerned with guilt, which has already been admitted, but with whose account of events to believe and thus the seriousness of the offence. In such cases the burden will be on the prosecution to prove its account to the usual standard i.e. beyond a reasonable doubt. The procedure in *Newton* will not arise when the facts in dispute are such that they would not materially affect the sentence imposed (*R v Bent* (1986)). Also, where a court hears an account which is wholly implausible or manifestly false it is entitled to reject that account without the need to hear evidence (*R v Hawkins* (1985)) . The court does not have to hear evidence in every case where there is a factual dispute before rejecting the version put forward by the defence (*R v Walton* (1987)).

## 10.9 DECISION AND SENTENCE

### 10.9.1 THE DECISION

Magistrates must find the accused person guilty or dismiss the matter. If the defendant is found guilty the court will often have a range of sentences that it may impose. Cavadino (1994)[4] observes that the Criminal Justice Act 1991, influenced by the concept of 'just deserts', emphasised greatly that magistrates courts should consider the seriousness of offences. In the preceding White Paper (Home Office, 1990, para. 2.2) it was stated that:

> Punishment in proportion to the seriousness of the crime has long been accepted as one of the many objectives in sentencing. It should be the principal focus for sentencing decisions.

Even after amendment of the Act the concept of seriousness is central to the decision-making process in the Magistrates Court. In the case of *R v Brown (Gary)* (1995) the Court of Appeal emphasised that courts should pass sentences appropriate to the seriousness of the offence committed and should not be deflected from that by the defendant's personal mitigation, feelings of guilt or depression. This emphasises the principle of 'just deserts' in the Criminal Justice Act 1991. Clearly personal mitigation can play a part in sentencing but only exceptionally should it reduce the level of sentence from that appropriate to the seriousness of the offence. Before passing sentence the accused or his legal representative must, however, be given an opportunity to speak in mitigation and address the court on sentence.

### 10.9.2 MITIGATION

Here it may be anticipated that there will be an emphasis on what wasn't said by the prosecution. This is done to indicate the offence was not as serious as it could be. For example, no damage was done by the incident, no person suffered food poisoning, no one's health was affected, no one was hurt by the occurrence. There will typically be statements made about how the guilty party has responded since the incident giving rise to the prosecution.

It may be said in mitigation that all similar machines now have guards fitted or the company has, since the incident, spent £5000 on arrestment plant or hygiene consultants have been brought in and all staff given further training on hygiene, or all repairs to the house have been completed. There will often be information put forward about the previous history of the defendant. It may be claimed that this is the first time such an incident has occurred in the 30 years that the shop has been open or the premises have been visited five times in the last six years and the Environmental Health Officer has never found any problems until this occasion.

Because many of the offences dealt with by Environmental Health Officers are strict liability offences it may be anticipated that any mitigation put forward will mention the fact that no fault attaches to the defendant. This invites the court to address difficult questions; for example, is a small amount of deliberate pollution to be viewed more or less seriously than a large amount of inadvertent pollution? Here any mitigation is clearly identifying that culpability may have little to do with guilt but is relevant to sentence.

In mitigation it may be also anticipated that the impact of the negative publicity attendant on conviction will be mentioned. This is particularly likely if the defendant is a company. In such cases it is usual to hear it said that the bad publicity will itself be a punishment and thus the court might wish to mitigate any further punishment in light of this. After hearing all of this and giving it due consideration the court will impose its sentence.

## 10.10  TRIAL ON INDICTMENT

It is possible, though relatively rare, that an Environmental Health Officer could find himself in a case before the Crown Court. Matters relating to this court will therefore only be dealt with in outline here. Self-evidently there are a number of differences between the Crown and Magistrates Courts. The most obvious is the presence of a judge and jury in the former and their absence in the latter. Essentially in a Crown Court the jury is concerned with matters of fact and the judge with matters of law. The judge, depending on the seriousness of the offence, is either a High Court judge, a Circuit judge, Recorder or an Assistant Recorder. The jurors will, in accordance with the Juries Act 1974, be persons between 18 and 70 years of age who appear on the electoral role and who have been resident in the UK for at least five years since the age of 13. The charges against a defendant will be contained in the indictment. The indictment is simply the printed accusation of the crime which is made at the suit of the Crown and read out at the start of the trial. It may contain one or more counts. Each matter will be put to the defendant who will plead guilty or not guilty to it. He may also plead guilty to a lesser offence. If the plea is guilty then the prosecution will present to the court a summary of the evidence in the case and the defence will put forward any mitigation. The defendant will then be sentenced. If the plea is one of guilty to a lesser offence, and this is accepted, the same will occur. If the plea of not guilty to a lesser offence is not accepted by the prosecution then the trial will continue.

### 10.10.1 JURY SELECTION

If unchallenged the juror takes an oath. This is in one of two forms:

> I swear by Almighty God that I will faithfully try the defendant and give a true verdict according to the evidence.

Or a juror may affirm:

> I do solemnly and sincerely and truly declare and affirm that I will faithfully try the defendant and give a true verdict according to the evidence.

After this the case will then proceed. Essentially it proceeds in a manner and order which is the same as that for a magistrates court. Thus the prosecution will make an opening speech giving an outline of the case against the defendant and then proceed to call his witnesses. These are subject to examination in chief, cross-examination and re-examination as before. The defence then proceeds in a similar way. Both sides have the right to make a closing speech. The time at which the prosecution is entitled to speak is after the close of the evidence for the defence and before the closing speech by or on behalf of the accused (Criminal Procedure (Right of Reply) Act 1964). The defence always gets the last word.

The judge will then sum up by telling the jury what the law is in the case and sum up the evidence on both sides. The jury will then retire to reach their verdict. This may be either:

(a) Guilty
(b) Not guilty
(c) Not guilty as charged but guilty of a lesser offence.

Assuming a guilty finding then, after any relevant reports, details of any previous convictions and any plea of mitigation the judge will pass sentence.

## 10.11 PENALTIES OF THE COURT

### 10.11.1 IMPRISONMENT

The maximum term of imprisonment which may be imposed for a single offence in a Magistrates Court is governed by s.31 of the Magistrates Court Act 1980 and limited to six months. This does not, however, permit the court to impose numerous six-month consecutive terms when a person is charged with more than one offence. Section 13 of the Magistrates

Court Act 1980 stipulates that a court shall not impose aggregate terms of imprisonment which exceed 12 months where two or more offences triable either way are tried summarily with the accused's consent. Nevertheless imprisonment for the Environmental Health offender is, and is likely to remain, a rare occurrence.

### 10.11.2 FINES

By far the most common penalty imposed by a Magistrates Court is the fine. The amount of fine which a court may impose is limited by the statutory maximum specified for the offence. The levels, which were amended by s.17 of the Criminal Justice Act 1991 are:

Level 1   £250
Level 2   £500
Level 3   £1000
Level 4   £2500
Level 5   £5000

An individual statute may provide for an increase on these maximum figures. For example offences under sections 7, 8 and 14 of the Food Safety Act 1990 can carry penalties when convicted on indictment of up to two years prison, an unlimited fine or both, or if convicted in the magistrates court six months prison or £20 000 fine or both. It would be wrong however to view the maximum penalty as the automatic tariff for an offence. The upper limit must be available but equally must only be appropriate for the most serious offence of the type in question. Therefore all other offences must fall somewhere below that maximum depending on how seriously the court views the offence. It is therefore inappropriate to enter the court anticipating the imposition of the maximum fine for an offence which is only typical (in the sense of average) of the type of offence in question.

### 10.11.3 DISCOUNT FOR A GUILTY PLEA

Section 48 of the Criminal Justice and Public Order Act 1994 has codified the common law position with regard to the duty of a court to consider giving a discount when an accused person pleads guilty. The Act provides that in determining what sentence to pass on an offender who has pleaded guilty to an offence in proceedings before it a court shall take into account:

(a) the **stage** in the proceedings for the offence at which the offender indicated his intention to plead guilty; and
(b) the **circumstances** in which this indication was given.

If as a result of taking into account any matters referred to above a court imposes a punishment on the offender which is less severe than the punishment it would otherwise have imposed, it must state in open court that it has done so. This is not as new as it might appear. The Magistrates' Association Sentencing Guidelines have for some time now advocated a reduction for a timely guilty plea. It is vital to note here that the two important elements which must be present are the timeliness of the plea and the circumstances surrounding it. There is no obligation to give a discount. In *R v Hollington* (1985) Lawton LJ put it this way:

> If a man is arrested and at once tells the police he is guilty and co-operates in the recovery of property and the identification of others concerned, he can expect a substantial discount. If a man is arrested in circumstances which he cannot hope to put forward a defence of not guilty, he cannot expect much by way of discount.

The courts have issued some guidance when a guilty plea is inevitable. In *R v Landey* [1996] the Court of Appeal found that the defendant had little choice but to plead guilty to aggravated vehicle taking when caught trapped in a car after a car chase. The court found that the maximum sentence of two years was appropriate, there being four previous convictions for the same offence. In *R v Hastings* [1996], again before the Court of Appeal, the defendant pleaded guilty to dangerous driving and driving with excess alcohol. He had been arrested at the scene of a collision following a chase by police. The court imposed the maximum two years sentence on the basis that the defendant had been caught in the worst possible case of dangerous driving.

Despite this guidance the Environmental Health Officer, who may have a low opinion of magistrates courts with regard to the level of sentence imposed, should understand the reasoning and rules behind some of the future decisions of the court and appreciate that some of the matters he puts before the courts will attract a discount.

### 10.12 COSTS

Where any person is convicted of an offence before a Magistrates or Crown Court then the court may make such order as to the costs that are to be paid by the accused to the prosecutor as it considers just and reasonable. Provision for Magistrates to award costs is contained in the Prosecution of Offences Act 1985, Part II and Regulations made thereunder (Costs in Criminal Cases (General ) Regulations 1986[6]).

The costs which are recoverable are those costs incurred after the decision to prosecute has been taken. In criminal cases the following orders for costs may be made:

1. When an information is not proceeded with or examining justices determine not to commit for trial or an information is dismissed on summary trial then defence costs may be from central funds.
2. Unless an order is made under s.17 of the Prosecution of Offences Act 1985 a private prosecutor's costs must be met by him. There is no power to order that a public authority, acting as a prosecutor, has its costs paid out of central funds. Here a public authority includes a local authority.
3. Under s.18 of the Prosecution of Offences Act 1985, if a prosecution is found to be 'just and reasonable' then costs may be paid by the accused on conviction and indeed an order should be made where the court is satisfied that the offender has the means and ability to pay (Practice Direction (1989)).
4. Under s.19 of the Prosecution of Offences Act 1985 the court can order either party to pay to the other costs incurred as a result of an unnecessary or improper act or omission by or on behalf of any party to the proceedings.

Environmental Health Officers may find themselves engaged in lengthy, time consuming and expensive investigations preparatory to the presentation of a case in the courts. Investigating officers should recognise the discretion the courts have to award costs in respect of the time they spend on investigation. In *Neville v. Gardner Merchant Ltd* (1984 ) it was found that where a Magistrates Court convicts a party and orders the party to pay the costs of the prosecution under the Costs in Criminal Cases Act 1973, s.2(2) (now repealed), the amount ordered to be paid may include an amount in respect of the time of the officer or person who investigated the alleged offence, notwithstanding that the officer is a salaried official of the prosecuting body.

The particular facts of the case concerned a number of offences under the Food Hygiene (General) Regulations 1970. The magistrates ordered the respondents to pay the costs of the prosecutor, including preparatory work carried out by the prosecutor's senior legal officer, but disallowed any amount in respect of time spent by the prosecutor's senior environmental health officer, who carried out the inspection which led to the proceedings. It was held that the justices had misdirected themselves in reaching the conclusion that they had no discretion to award the costs which had been disallowed. Not only did they have the discretion to award the costs which had been disallowed, but prima facie costs of this kind should be awarded. The fact that the costs related to the time of an investigating officer paid out of public funds whose duty it was to investigate such offences did not preclude the award of costs.

If a court finds, in the words of Lord Lane CJ in *Tottenham Justices ex parte Joshi* (1982), that all the time and trouble of the investigating officer had been due to the offences committed by this respondent, then it would be right to award the whole of these costs.

The Court of Appeal has in *R v Associated Octel Co. Ltd (Costs)* 1996 reconsidered *Neville v Gardner Merchant* when costs of £142 655.33 were being claimed by the Health and Safety Executive. The court again found that the costs of prosecution could include the costs of the prosecuting authority in carrying out investigations with a view to the prosecution of a defendant who was then convicted. The court also went on to usefully observe that:

- The prosecution should serve upon the defence full details of its costs at the earliest time so as to give the defence a proper opportunity to consider them and make representations if appropriate; and
- Once served with a schedule of the prosecution's costs, a defendant, if he wished to dispute the whole or any part of it, should give proper notice to the prosecution of any objections he proposed to make, or at least make it plain to the court what those objections were.

Thus where a local authority is claiming costs these must be specified when making the claim. If the court decides that the amount claimed is just and reasonable it may order the accused to pay such costs. It is important therefore that it is recognised how vital it is that Environmental Health Officers specify all costs. These may include legal fees, analyst and expert fees, investigative and administrative costs. A typical costs specification may be as shown in form 10.1.

### 10.13  DISQUALIFICATION OF DIRECTORS

Section 2 of the Company Directors' Disqualification Act 1986 provides that a court may by order disqualify any persons convicted of indictable offences connected with the promotion, formation, management or liquidation of a company. Such an order when made will prohibit an individual, without leave of the court, from being:

(a)  A director of a company
(b)  The liquidator or administrator of a company
(c)  The receiver or manager of a company's property
(d)  Concerned in any way, directly or indirectly, with, or from taking part in, the promotion, formation or management of a company,

for a specified period not exceeding five years beginning with the date of the order.

Here the term 'management' has been found to extend to the management of health and safety in the workplace. For example in June 1992, Lewes Crown Court disqualified a company director following a prose-

## COSTS

**Anytown City Council v** ..........................................
**Before** ......................... **Court on the** ............ **day of** ................... **19**.........

### Investigative and Preparatory Costs

| | | |
|---|---|---|
| Officer Time | .....hr at .......hr | = £ ........ |
| Administrative Time | ......hr at ......hr | = £......... |
| Exhibits | | = £......... |
| Travel | ....... miles at £......./mile | = £........ |
| | | = £......... |
| Other (Specify)..................................... | | = £.......... |

### Cost Of Witness Attendance at Court

| | | |
|---|---|---|
| Officer attendance | ........hr..... at...../hr | = £......... |
| Travel | ........hr..... at...../hr | = £.......... |
| Other Witnesses.................................................... | | |

### Expert Witness

| | | |
|---|---|---|
| Expert Witness | ........hr...... at...../hr | = £......... |
| Travel | ........hr...... at...../hr | = £......... |
| Analyst................................................................................. = £.......... | | |

### Legal Costs

| | | |
|---|---|---|
| Legal fee claimed on behalf of ....... | | = £.......... |

**Total Costs**                                    **£...........**

Form 10.1 Costs

cution for contravention of a Prohibition Notice under the Health and Safety at Work etc. Act 1974. There is no reason however why the power of disqualification should be limited only to matters of health and safety. In *R v Corbin* (1984) the management of the company was held to refer to both the internal and external affairs of that company. As one aim of the 1986 Act is to protect the public and since environmental health law and its enforcement is clearly about public protection the applicability of this provision to the work of the Environmental Health Officer is clear. In certain instances it would be appropriate for those involved in prosecution to consider the option of seeking the disqualification of one or more company directors. In doing so it should be borne in mind that it is not considered sensible to seek to prohibit an individual from engaging in business activities and at the same time to seek a compensation order which is likely to depend on such activities for payment (*R v Holmes* [1991]).

### 10.14 COMPENSATION ORDERS

The criminal courts, though not generally concerned with the question of compensation for loss or damage sustained by a victim of a criminal offence, do have the power to make awards of compensation. Section 35 of the Powers of Criminal Courts Act 1973 empowers a criminal court to order any person convicted of a criminal offence to pay compensation to the victim in addition to any other sentence the court may impose. Guidance on the use of compensation orders may be found in Home Office Circular No. 53/1993: Compensation in the Criminal Courts.

Section 35(1) of the Powers of the Criminal Courts Act 1973 states;

> ... a court by or before which a person is convicted of an offence, in addition to dealing with him in any other way, may, on application or otherwise make an order (in this Act referred to as a 'compensation order') requiring him to pay compensation for any personal injury, loss or damage resulting from that offence which is taken into consideration by the court in determining sentence.

Under section 67 of the Criminal Justice Act 1982 courts are empowered to order payment of compensation either instead of, or in addition to, dealing with the offender in any other way. A compensation order may therefore be a sentence in its own right. The Criminal Justice Act 1988 (s.104(1)) amended section 35 (Powers of Criminal Courts Act 1973, substituted by Criminal Justice Act 1982 s.67) to emphasise the requirement for courts to consider compensation in every case involving personal injury, loss or damage and provided that a court must give reasons if a compensation order is not made.

'A court shall give reasons on passing sentence if it does not make [a compensation] order in a case where this section empowers it to do so.' Reasons must be given in open court and, in magistrates courts, recorded in the register (Magistrates' Courts Rules 1981 rule 66 (10 A)). More recently the Criminal Justice Act 1991 raised the maximum compensation order which a magistrates court may impose for any one offence from £2,000 to £5,000 (Criminal Justice Act 1991, s.17(3)(a) and Sch.4, Part I by amendment of Magistrates Court Act 1980, s.40(1)). The courts have established that a compensation order is only suitable for the most straightforward cases and is intended as a speedy way of dealing with simple cases in which no great amounts are at stake. In *Herbert v Lambeth* LBC (1991) information was laid under s.99 of the Public Health Act 1936 (now s.82 Environmental Protection Act 1990) alleging Lambeth LBC were responsible for a statutory nuisance as defined by s.92 of that Act (now s.79 Environmental Protection Act 1990). Magistrates found the London Borough of Lambeth liable under s.94(2) of the 1936 Act and made a Nuisance Order requiring works to be carried out to abate the Statutory Nuisance. The magistrates however refused to make a compensation order under s.35 of the Powers of the Criminal Courts Act 1973. Lord Justice Woolf sought to underline what should be the approach of magistrates with regard to compensation and referred to *Stone's Justices Manual* (123rd ed. (1991) para. 3/785, p.843) which stated:

> The machinery of a compensation order under this Act is intended for clear and simple cases. It must always be remembered that the civil rights of the victim remain, although the power to make a compensation order is not confined to cases where there is a civil liability. A compensation order made by the court of trial can be extremely beneficial as long as it is confined to simple straightforward cases and generally cases where no great amount was at stake.

Woolf considered it would not be appropriate to use the powers in section 35 to award substantial compensation sums for matters which could loosely be described as personal injury. Where a civil action may be brought, that was the proper and preferable way to deal with any award of a substantial amount. Guidance on the use of compensation orders is to be found in the case of *R v Miller* (1976). Here the court restated the general principles to be followed when making an order.

1.  First and foremost, the order should be made only where the legal position is clear.

2. The order must be precisely drawn and must be related to the offence of which the offender has been convicted or which he has asked to have taken into consideration on sentence.
3. The court must have regard to the offenders means.
4. The order must not be oppressive.

Circular 53/93 makes it clear that compensation may be ordered for distress and anxiety caused by threats for example, and for mental as well as physical injury (*Bond v Chief Constable of Kent* [1983]). The assessment of compensation in such cases is recognised as often difficult, but some of the factors which may be taken into account are any medical or other help required, the length of any absence from work and a comparison with the suggested levels of compensation for physical injury. The annex to Circular 53/93 updates guidance on compensation in personal injury cases. The approach adopted is based on awards currently made by the Criminal Injuries Compensation Board. The Board bases its awards on common law damages. Guidance on the assessment of common law damages for a wider range of more serious injuries can be found in a recent publication by the Judicial Studies Board, *Guidelines on the Assessment of General Damages in Personal Injury Cases.*

### 10.15 TRIBUNALS

English courts are often seen as overly bound by precedent, rigid, formal and expensive. These features are often said to limit the accessibility of the courts and thereby limit their usefulness. Judges and Magistrates are not always seen as the ideal arbiters, and practitioners have expressed their doubts about the suitability of the criminal process when dealing with environmental matters. Indeed at least one Lord Justice of Appeal has stated that he does '. . . have reservations as to whether the criminal courts are the appropriate tribunal to determine some of the offences created by environmental legislation'.[7]

Some writers go further and even call for the establishments of 'Environmental Courts', perhaps modelled on the South Australian model,[8] or in the words of Carnwath (1992)[9] a lower tier court operating at a lower level requiring a rationalisation of the existing jurisdictions of Magistrates, County and Crown Courts. As yet such courts exist only in the thoughts of writers and commentators. There does exist however an alternative forum for the resolution of some legal disputes. These are the Tribunals, which cover such areas as Immigration, Social Security and Industrial Relations, and are employed to deal with a large number of the less serious or more technical legal disputes. Tribunals in many ways resemble courts. They are a judicial body able to reach a binding decision and are independent of the administration.

### 10.15.1 HISTORY

In 1955 a Committee of Administrative Tribunals and Inquiries was set up to investigate tribunals and inquiries generally.[10] Under the chairmanship of Sir Oliver Franks this committee investigated the constitution and workings of tribunals and inquiries. The Franks Report of 1957 resulted in the Tribunals and Inquiries Act 1958, which was later consolidated by the Tribunals and Inquiries Acts of 1971 and 1992. A recommendation of the Committee was the establishment of the Council on Tribunals which has as its role the oversight of the operation of the system. Section 1 of the 1971 Act stipulated the functions of the Council, which were to keep under review the constitution and working of the tribunals and from time to time report on their constitution and working. The Council was also to consider and report on such matters as may be referred to the Council.

## 10.16 INDUSTRIAL TRIBUNALS

There are very many tribunals, however the one the Environmental Health Officer is most likely to encounter is the Industrial Tribunal, here sitting in an appellate capacity to hear appeals against notices issued under s.21 of the Health and Safety at Work etc. Act 1974 (Improvement Notices) and under s.22 of the same Act (Prohibition Notices).

These tribunals were originally established under s.12 of the Industrial Training Act 1964. However since their inception their jurisdiction has increased and now extends to the Industrial Relations Act 1971, the Equal Pay Act 1970, the Sex Discrimination Acts 1975 and 1986, the Health and Safety at Work etc. Act 1974 and the Employment Acts 1980, 1982, 1988, 1989 and 1990.

### 10.16.1 COMPOSITION

There are three members to a tribunal: a legally qualified Chairman plus two others. The tribunal will generally be less formal than a court of law and for this reason legal representation is not required. However, in practice, it is very common for parties to be represented by someone of their choice. Hearings are generally open to the public. Proceedings are regulated according to the rules of the tribunal applicable to the matters in dispute. For a tribunal hearing an appeal under the Health and Safety at Work etc. Act 1974 the procedure is regulated by the Industrial Tribunals (Constitution and Rules of Procedure) Regulations 1993.

10.16.2 THE INDUSTRIAL TRIBUNALS (CONSTITUTION AND RULES OF
PROCEDURE) REGULATIONS 1993

These regulations replace the earlier Industrial Tribunals (Improvement
and Prohibition Notices Appeals) Regulations 1974 and set out the rules
of procedure of industrial tribunals in England and Wales for the deter-
mination of appeals against improvement and prohibition notices issued
under the Health and Safety at Work etc. Act 1974.

The fourth schedule to the regulations details the Rules of Procedure
and covers matters relating to:

- The time limit for bringing an appeal
- Action upon receipt of an appeal
- The power to require attendance of witnesses
- The time and place of hearing and the appointment of an assessor
- The hearing
- The decision of the tribunal and the review of that decision
- Costs.


10.16.3 APPEALS

Anyone receiving either of the two types of notice can appeal (s.24(2)
Health and Safety at Work etc. Act 1974) to an industrial tribunal on one
of several grounds:

1. The breach of law is argued as *de minimis*
2. Wrong interpretation of the law
3. The officer serving the notice has acted *ultra vires*
4. Breach of law has occurred but the proposed solution is not practicable or reasonably practicable.

On hearing an appeal the tribunal may affirm, modify or cancel the
notice. If it affirms it, it may do so with modifications ('modifications' are
defined in s.82(1)(c) of the Act), i.e. additions, omissions or amendments
(s.24(2)). The effect of an appeal is detailed in s.24(3) of the Health and
Safety at Work etc. Act 1974. In the case of an improvement notice, an
appeal has the effect of suspending the operation of the notice. In the
case of a prohibition notice an appeal only suspends the operation of the
notice if the tribunal so directs and the suspension is effective from the
time the tribunal gives this direction. A tribunal is not permitted to modify a notice so as to introduce a duty to comply with a requirement not
mentioned in the original form of the notice. Any evidence put before
the tribunal may be subject to evaluation by expert assessors who may be
brought in at the discretion of the tribunal (s.24(4)). After hearing the
appeal the tribunal may cancel or confirm the notice. Confirmation can

result in the notice being left unchanged or confirmation may be of an amended form of the notice. A tribunal may make amendments to notices, which may include adding to the requirements of the notice (*Tesco Stores v Edwards* [1977]) but this does not extend to allowing it to introduce a duty to comply with any statutory provision previously omitted from the notice. In *British Airways Board v Henderson* [1979] an industrial tribunal held that its power to modify only related to the breaches of a statute referred to in the notice; it could not allow the request of the Environmental Health Officer to amend the notice stipulating a breach of s.3 of the Act in addition to the breach of s.2 already contained therein.

### 10.16.4 PROCEDURE

An appeal to an industrial tribunal against an improvement or prohibition notice is commenced by the appellant sending to the Secretary of the Office of the Tribunals a written notice of appeal. The notice of appeal must be sent to the Secretary within 21 days from the date of the service of the notice appealed against. A tribunal may extend the time where it is satisfied, on an application made in writing to the Secretary, that it is not or was not reasonably practicable for an appeal to be brought within 21 days.

A notice of appeal should contain:

(a) the name and address of the appellant and, if it is different, an address within the United Kingdom to which he requires notices and documents relating to the appeal to be sent;
(b) the date of the improvement notice or prohibition notice appealed against and the address of the premises or place concerned;
(c) the name and address of the respondent;
(d) particulars of the requirements or directions appealed against; and
(e) the grounds of the appeal.

An example of such an appeal is reproduced here (form 10.2).

Upon receipt of the notice of appeal the Secretary is obliged to enter particulars of it in the Register and must send a copy of it to the respondent. He must inform the parties in writing of the case number of the appeal entered in the Register and of the address to which notices and other communications to the Secretary shall be sent. The case number from this point on constitutes the title of the proceedings.

In the case of an appeal against a prohibition notice in which an application is made under s.24(3)(b) of the Health and Safety at Work etc. Act 1974 to suspend the operation of the notice until the appeal is finally disposed of or withdrawn, an application must be sent in writing to the Secretary and must set out:

## APPEAL TO AN INDUSTRIAL TRIBUNAL UNDER THE HEALTH AND SAFETY AT WORK ETC. ACT 1974

TO: The Secretary of the Tribunals
Central Office of the Industrial Tribunals
Southgate Street
Bury-St-Edmunds
IT33 2AQ

**1. Full name of Appellant (or title if company or organisation):**

Slip Slide Company Limited

**2. Address of Appellant (registered office if applicable)**

12 Sunnyside Road, Anytown, AN1 1AB

**3. Address of Appellant or his representative for service of documents if different from 2 above**

J. Smith Tel. No. 0111 123456

**4. Details of Notice appealed against:**

Improvement Notice dated 17th April 1997, Serial No. 123/4567

**5. Address of the premises or place to which the Notice refers:**

12 Sunnyside Road, Anytown, AN1 1AB

**6. Name and full address of Inspector as shown on Notice:**

Mabel Margaret Mabeline, Anytown District Council Environmental Health Dept, Town Hall Anytown, AN1 2AB

**7. Particulars of the requirements or directions appealed against:**

Paragraphs 1, 2 & 3 of the schedule to the Improvement Notice require works which are not reasonably practicable.

Signed ..... *I M Lawyer* ............

Dated ....20 April 1997........

**Form 10.2** Appeal to an industrial tribunal under the Health and Safety at Work etc. Act 1974

(a) the case number of the appeal if known to the appellant or particulars sufficient to identify the appeal; and

(b) the grounds on which the application is made.

On receipt of the application, the Secretary must enter particulars of it against the entry in the Register relating to the appeal and must send a copy of it to the respondent. The tribunal has the power to require attendance of witnesses and production of documents etc.

On the application of a party to the appeal, made either by notice to the Secretary or at the hearing, the tribunal may:

(a) require a party to furnish in writing to another party further particulars of the grounds on which he relies and of any facts and contentions relevant thereto;

(b) grant to a party such discovery or inspection of documents as might be granted by a county court; and

(c) require the attendance of any person as a witness or require the production of any document relating to the matter to be determined; and

(d) may appoint the time at or within which or the place at which any act required in pursuance of this requirement is to be done.

Any person on whom such a requirement has been made is permitted to apply to the tribunal either by notice to the Secretary or at the hearing to vary or set aside the requirement.

Not less than 14 days before the date fixed for the hearing (or such shorter time as is agreed by the Secretary with the parties) the Secretary must send to each party a notice of hearing together with information and guidance as to:

- attendance at the hearing
- witnesses
- the bringing of documents
- representation by another person
- written representations.

The tribunal may if it thinks fit postpone the day or time fixed for, or adjourn, any hearing. Normally the hearing of an appeal will take place in public, though there are circumstances where the tribunal will not sit in public. These will be:

1. Where a Minister of the Crown has directed a tribunal to sit in private on grounds of national security.

2. If on the application of a party the tribunal considers it appropriate to do so, for the purpose of hearing evidence;

(a) which relates to matters of such a nature that it would be against the interests of national security to allow the evidence to be given in public, or

(b) it would be hearing evidence from any person which in the opinion of the tribunal is likely to consist of information the disclosure of which would cause substantial injury to the undertaking of the appellant or of any undertaking in which he works.

### 10.16.5 PROCEDURE AT THE HEARING

If either party to the appeal fails to appear or to be represented at the time and place fixed for the hearing, the tribunal may adjourn the hearing to a later date or it is empowered to deal with the appeal in the absence of that party. Before so disposing of the matter under such circumstances the tribunal is obliged to consider any written representations submitted by the missing party.

At the hearing, as in a trial, the parties are entitled to make an opening statement, to give evidence, to call witnesses, to cross-examine any witnesses called by the other party and to address the tribunal. Any witnesses called may, at the discretion of the tribunal, be required to give their evidence on oath or affirmation.

In reaching a decision a tribunal composed of three members may do so by a majority. If a tribunal is composed of two members only then will the chairman have a second or casting vote. The decision of the tribunal is recorded in a document signed by the chairman which also contains the reasons for the decision. The person appointed as clerk to the tribunal by the Secretary is obliged to transmit the document, signed by the chairman, to the Secretary who in turn must enter it in the Register of applications, appeals and decisions. This register has to be kept in accordance with regulation 9 of the Industrial Tribunals (Constitution and Rules of Procedure) Regulations 1993. The Secretary must also send a copy of the entry to each of the parties. The Register is open to public inspection without charge.

A tribunal has the power on the application of a party to review and revoke or vary by certificate under the chairman's hand any of its decisions on the grounds that:

(a) the decision was wrongly made as a result of an error on the part of the tribunal staff;

(b) a party did not receive notice of the proceedings leading to the decision;

(c) the decision was made in the absence of a party;

(d) new evidence has become available since the making of the decision provided that its existence could not have been reasonably known of or foreseen; or

(e) the interests of justice require such a review.

If the application for review is not made at the hearing, then such application must be made to the Secretary within 14 days from the date of the entry of a decision in the Register and must be in writing stating the grounds in full. Such an application may be refused by the chairman of the tribunal which decided the case, by the President of the Industrial Tribunals (England and Wales) or by a Regional Chairman if in their opinion the application has no reasonable prospect of success. In such a case whoever makes the decision must state the reasons for this opinion. Where an application is accepted, it must be heard by the tribunal and if it is granted the tribunal is required to either vary or revoke its decision and order a rehearing. Before granting an application the party making the application may at the discretion of the tribunal be required to give notice thereof to the other party.

10.16.6 COSTS

The tribunal has the power to award costs and may make an order that one party is to pay to the other either a specified sum in respect of the costs incurred in connection with the appeal or, in default of agreement, the taxed amount of those costs.

**NOTES**

1 These are reproduced in *The sentence of the Court; A handbook for Magistrates*, Watkins M, Gordon W, Jeffries A, 1995, Waterside Press, Winchester.

2 In January 1995 there were 25 723 Magistrates in England and Wales and 4365 in the Duchy of Lancaster. Sixty-four are stipendiary magistrates.

3 Lord Mackay of Clashfern (1994), *The Administration of Justice*, London, The Hamlyn Trust, Stevens & Sons/Sweet & Maxwell.

4 Lord Mackay of Clashfern (1994, *Op cit*). To merit appointment potential magistrates must reside in or within 15 miles of the area of the bench (Justices of the Peace Act 1979, s.7), and thus be part of, and drawn from, the community which they serve.

5 Cavadino P, Clayton A, Wiles P, 1994, *Seriousness of Offences: The Results of the South Yorkshire Study*, Joint Board for Research & Development in Criminal Justice.

6 See also: *Guide to the Awards of Costs to Criminal Proceedings*, HMSO, ISBN No. 0/11/380048/7.

7 Lord Justice Woolf, 4th Annual Garner Lecture, Are the Judiciary Environmentally Myopic? (1992) *Journal of Environmental Law*, Vol. 4, No. 1. p.10.

8 Upton W, 1994, Environmental Courts – The South Australian Initiative, *Environmental Law*, Vol. 8, Nos. 1 & 2, 12, Spring/Summer 1994. McAuslan P, 1991, *Op. cit*.

9 Carnwarth R (1992), Environmental Enforcement: The Need for a Specialist Court, *Jnl Planning and Env Law*, Sept., 799.

10 Franks Committee, Cmnd. 218 (1957).

# Table of cases

# Table of Statutes

# Statutory Instruments

# Index